Deadly Turbulence

Deadly Turbulence

The Air Safety Lessons of Braniff Flight 250 and Other Airliners, 1959–1966

Steve Pollock

McFarland & Company, Inc., Publishers
Jefferson, North Carolina

LIBRARY OF CONGRESS CATALOGUING-IN-PUBLICATION DATA

Pollock, Steve, 1963– author.
    Deadly turbulence : the air safety lessons of Braniff flight 250 and other airliners, 1959–1966 / Steve Pollock.
      p.   cm.
    Includes bibliographical references and index.

    ISBN 978-0-7864-7433-2 (softcover : acid free paper) ∞
    ISBN 978-1-4766-1326-0 (ebook)

    1. Aircraft accidents—United States—Case studies. 2. Jet transports—Accidents—United States. 3. Jet transports—United States—Safety measures. I. Title.

TL553.52.P65 2014
363.12'465—dc23   2014003862

BRITISH LIBRARY CATALOGUING DATA ARE AVAILABLE

© 2014 Steve Pollock. All rights reserved

No part of this book may be reproduced or transmitted in any form or by any means, electronic or mechanical, including photocopying or recording, or by any information storage and retrieval system, without permission in writing from the publisher.

On the cover: Braniff's N1553, 1966 (courtesy Bob Proctor, Jon Proctor Collection); *Falls City Journal* photograph of the wreckage (© 2014 Bill Schock, used by permission)

Manufactured in the United States of America

McFarland & Company, Inc., Publishers
Box 611, Jefferson, North Carolina 28640
www.mcfarlandpub.com

For Frank

Table of Contents

Acknowledgments ix
Preface 1
Prologue 5

Part I. "I don't sleep well nights"

1. The Day the Movies Almost Died: Pan American World Airways Flight 115 9
2. Melting Into a Threatening Sky: Northwest Orient Airlines Flight 705 19
3. "Unusual conditions and circumstances": The Investigation of Northwest Orient Flight 705 28
4. "A turbulence problem"? Persistent Questions 44
5. "It broke the sound barrier at least once": United Airlines Flight 746 47
6. "The Nation's Most Progressive Airline": Eastern Air Lines 52
7. A Strange, Weird Thing: Eastern Air Lines Flight 301 59
8. Disappeared into the Overcast: Eastern Air Lines Flight 304 ... 64
9. Extraordinary Lives: Grant Newby, Marie-Hélène Lefaucheux and Kenneth Lee Spencer 69
10. "Extreme disintegration of the aircraft structure": The Investigation of Eastern Air Lines Flight 304 82

Part II. Falls City

11. Dusting Off a Dusty Little Airline: Braniff International Airways 99
12. Airstrip: "The End of the Plain Plane" 107
13. Fastback Jet: The BAC One-11 113
14. Happy Birthday! Bill Schock and That Night in August 124
15. Milk Run: Braniff International Flight 250 128
16. "One big, terrible nightmare": Flight 250 and Falls City 134
17. "Ease power back": The Investigation of Braniff Flight 250 160
18. Mr. Tornado: Dr. Ted Fujita and the Squall Line of August 6–7, 1966 169
19. Knocked from the Sky: A High-Intensity Force 179

Part III. Postscript

20. New Normals: The "Before" and "After" Times 187

Notes 195
Bibliography 203
Index 213

Acknowledgments

I appreciate readers for taking time to read some forgotten stories which played important roles in ensuring that air transport, a crucial component of our modern economic life, is as safe as possible.

No author is an island, no book produced in isolation; this book is no exception. Here is my chance to thank a few of the people who directly or indirectly helped make the book a reality.

Although space and other limitations means a full, exhaustive listing of influences on this project is impossible, I am forever indebted to the following for their gracious assistance:

Falls City, Nebraska

Thanks first and foremost go to Bill Schock, newspaper owner, editor, columnist and reporter extraordinaire; B-17 pilot of the Eighth Air Force with 25 missions to his credit; prisoner of war; one of the first to arrive at the crash site of Braniff Flight 250; and generous gentleman of the old schools of life and journalism. Without him, this book would have been impossible. Special thanks are due him for so graciously granting permission to publish photos and quote from articles. Thanks are due also to Bill's son and grandson, Scott and Jason Schock, second and third generation powers behind the *Falls City Journal*. The majority of the photos herein are selected from those taken by Bill Schock in the hours and days following the crash of Braniff 250. These are heart-wrenching and fascinating and are invaluable in understanding the Braniff 250 experience. I am grateful to the Schocks and the *Falls City Journal* for permission to print them.

Thanks also go to Falls City's mayor, Tim Hersh, and his wife and family for taking time to talk about their town, past, present and future. They helped provide needed perspective on how the town's past relates to its future, shared their passion, and officially welcomed me to town.

Thanks go to Kenny and Elaine Schawang for taking time out of their busy schedules to meet at their farm and spend time at the Braniff 250 crash site. They also extended the privilege of sitting in Vernell Schawang's old kitchen, which helped me visualize that dark, violent night that was so devastating for her family. Kenny and Elaine helped fill in details of that long-ago nightmare for yet another nosy stranger; they also constructed and meticulously maintain the memorial to Flight 250 which is pictured herein.

Richardson, Texas

Working with the staff at the University of Texas-Dallas, Richardson Campus, location of the Braniff Collection within the History of Aviation Collection of the McDermott Library, was a great experience and uncovered many of the inner workings of an airline responding to the loss of its workers and passengers. Curator Patrizia Nava during visits in June and October and curator Dr. Thomas J. Allen, during a later October visit, made the research process quick, efficient and stress-free. Thanks are also due to Pat Zahrt, Braniff public relations officer and editor of the company newsletter, the *B-Liner*. When the 1982 shutdown occurred, she and others made sure that the PR files at least were taken to UTD for safe-keeping, meaning future researchers would have important sources for their work.

Lubbock, Texas

The staff of the Southwest Collection/Special Collections Library of Texas Tech University in Lubbock was wonderfully helpful during the research into Dr. Tetsuya Fujita's archives. Manager of Wind Science and Engineering Research Center Carolann Stanley and Special Collections Archivist Dr. Monte Monroe helped set up the research and provide necessary materials. Thanks are due to Dr. Fujita's family for making his archives centrally located and accessible.

Boulder, Colorado

Dr. Fernando Caracena, who was willing to work with me on the book, is always insightful. While the book grew too large to include the later downburst accidents which Dr. Caracena and Dr. Fujita worked on, Dr. Caracena's wisdom and memory formed an invaluable foundation.

Airliner Photographers

The willingness of classic airliner photographers who shared some of their old, rare shots of aircraft is appreciated. Jon Proctor of Sand Point, Idaho, provided one of only three known photos of the Braniff Flight 250 aircraft, BAC One-11 N1553, from his father's collection. Eddie Coates of Raleigh, North Carolina, permitted the inclusion of the aircraft which crashed while being flown on Eastern flight 304. And David L. Schulman granted the opportunity to use his fascinating photo of an Eastern Air Lines DC-8-21 on the ramp at Miami for the book as well. The openness of all three to share their work was integral.

Eden Prairie, Minnesota, and Smyrna, Georgia

Tim Hilliker, son of Braniff Flight 250's first officer Jim Hilliker, was instrumental in sparking this project into life. A wonderful man with a clear picture of his father, Tim is always full of ideas and encouragement and provided a crucial boost at critical points to get the story out there.

Tim's brother Dan and sister Kathy and their Uncle Bob Lindseth made great contributions and a nice welcome into Dan's home and opened up about what they experienced on that terrible night all those years ago. A great debt is owed them for being willing to share that pain so that others can know that progress often comes at a price.

On a personal level, family and friends, especially my parents, Marion and Janis Pollock, and sisters Kathy and Vicky and their extensive families, were enthusiastic and supportive of this idea. "You can do it if you set your mind to it," is my parents' mantra, and all of us benefit from their lessons in quiet determination. Close friends Stan Bedford, Jay McGinnis, Don P. Brown and his wife Jean have given support and encouragement generously for many years. Close friend David Garms is appreciated as well; research trips wouldn't have been possible without his support. Fellow reporters at the *Duncan Banner* Glen Seeber and Linda Walker-Speers taught me much about daily small-town journalism.

My partner, Frank Lester, has put up with me for more than 14 years and been my champion and defender. I am most grateful of all to Frank for sharing the ride in life and work. His serious support for my effort is what made that effort possible. His understanding, concern and love are the oxygen I breathe, and his library knowledge, editing skills and rewrite suggestions were invaluable. For this and many other things, I am most grateful to him.

Preface

Man's flight through life is sustained by the power of his knowledge.
—Austin "Dusty" Miller, 1958

In the main, this is mostly the story of three commercial airline accidents, their victims, and what happened next. This is the story of the tangible and intangible impact on friends and families of the victims of such crashes as Braniff International Airways Flight 250. But the story is not confined to their loss. Flight 250 affected commercial aviation technology and operations, and was a stepping-stone in the development of greater understanding in the field of meso-meteorology and safer flight. In fact, Flight 250 was one of several important stepping-stones over a long period of developing a safer American commercial aviation industry. But it was not the first, nor would it be the last, crash during turbulent weather. All of these crashes were bittersweet, causing huge destruction on the one hand, but building important human understandings on the other.

That legacy is the story here. Why else should we pay attention today to rather obscure and mostly forgotten plane crashes from a half-century ago? Braniff Flight 250 would be remarkable for its human impact alone, which is pretty much true of any of the crashes in commercial aviation's first hundred years. But the scientific and technical investigation into Flight 250's demise saw the first use of cockpit voice recorder technology in an aviation accident investigation and then resulted in important understandings about the relation of jet aircraft to the often hostile environment in which they operate.

In fact, study of the hostile environment Flight 250 attempted to operate in was an important stepping-stone itself. For a remarkable man named Tetsuya "Ted" Fujita, called in by British Aircraft to conduct a more intensive investigation of the weather giant which batted their carefully designed and built jet from the August night sky, Flight 250 represented an important opportunity to further develop the crucial study of meso-meteorology. Five years later, Dr. Fujita would use the knowledge gained from extensive study of "Tornado Alley" storms and the damage they produce, including the data and knowledge from Flight 250, to create the Fujita Scale of Tornado Intensity, the same F-scale that we hear about so frequently during storm season. Braniff 250 was the first time Dr. Fujita investigated the effects of turbulence on commercial aviation, and the knowledge he gained would be called upon several more times. The investigations into the crashes of Eastern Flight 66 at New York's John F. Kennedy International Airport in 1975, Pan American World Airways Flight 759 at Kenner, Louisiana, in 1982, Delta Air Lines Flight 191 at Dallas-Fort Worth International Airport in 1985, and USAir Flight 1016 at Charlotte, North Carolina, in 1994, were all built on what was learned on August 6, 1966.

While we should never forget the human toll of any accident, we should also never forget to make "lemonade of lemons," so to speak. While captain Donald Pauly, first officer James Hilliker, hostesses Ginger Brisbane and Sharon Hendricks, and their 38 passengers died that stormy August night as Braniff 250 pancaked into Antone Schawang's soybean field, it is a certainty that their deaths were far from meaningless.

As of this writing in 2013, the United States has not lost an airliner in turbulence or microburst/wind shear since USAir Flight 1016 in Charlotte in 1994. It's an enviable record. The old saying applies that you are in far more danger driving to the airport, or even taking a bath in your own tub at home, than you are sitting in a metal tube shooting along at 500 mph at 35,000 feet. You're also probably more at risk of death from eating that stale pizza you picked up at the airport kiosk than you are from your airliner crashing. As Les Lautman, a safety manager of the Boeing Commercial Aircraft Company, put it in a more scientific and statistical way in 1989, "If you were born on an airliner in the U.S. in this decade and never got off you would encounter your first fatal accident when you were 2,300 years of age and you would still have a 29 percent chance of being one of the survivors."[1]

To the average Joe, statistics are funny things. But crashes with charred, broken airplanes and shrouded dead bodies are very real. As my own mother would put it, "It only takes once." Something deep within us, perhaps a prehistoric instinct left over from battling our environment, tells us that the danger is in sitting in a metal tube shooting along at 500 mph at 35,000 feet, not in getting out of our own bathtubs. Perhaps contributing to our "gut feeling" is that airliner crashes receive prominent play on newspaper front pages, Web sites and television news programs, which ignore the death of an elderly man in Dubuque after he slips and falls while taking his nightly bath. Even though such home-related injuries result in nearly 20,000 deaths and 21 million medical visits on average each year[2] (says the Washington, D.C., Home Safety Council, a "national nonprofit organization solely dedicated to preventing home-related injuries"), they don't get nearly the press that a good, juicy airliner crash does. What steps did we take in order to get to the point where our bathtubs are more dangerous than a 900,000-pound Boeing 747–8?

The history of commercial aviation safety is long and complex, as well as outside the scope of this book and the expertise of the writer. As a journalist, I can, however, tailor a more specific question and ask, "Why did turbulence crash airliners during the early Jet Age and why doesn't it happen now?" The simple, short answer is that (a) we didn't know much about the safe operation of a jetliner in turbulence because jetliners didn't exist in service until 1959, and (b) it took several major crashes such as the ones here described, as well as other more minor incidents and the loss of many lives and millions of dollars and years of investigation, to produce our record of safety in a hostile environment. We can never stop learning and improving as long as people want to fly from Dubuque to Cape Town. People gave their lives and we reap the benefits. We should be paying attention.

Prologue

*A night of menace, touched and tainted by
an evil wind. A difficult night to conquer.*
—Antoine de Saint-Exupéry

---

*Northeast of Falls City, Nebraska
11:12 p.m., August 6, 1966
Saturday*

A brilliant, full, round moon hung in the eastern sky, lighting up the southeast Nebraska prairie. Darkened houses and white cows were illuminated in a ghostly palette by the lunar brightness. The cows ate hay in front of deserted gray barns and glanced up as a car rumbled past. The mood in the car was light, easy and fun. The passengers laughed at how the moon was ducking in and out behind the clouds. The car churned up a long, expanding cone of dust, which hung in the hot August air, swirling and curling in the car's wake. The dust settled slowly on frogs croaking loudly in the ditches, and the air was warm and humid. His shirt collar stuck uncomfortably to the driver's neck, which was stained a deep red/brown from successive long summers of tending crops. A bottle of pop at home sounded like the best thing in the world, and he thought about the moment when he would draw it from the cooler, bits of crushed ice clinging to the bottle and dripping down onto the porch floor. For the farmer and his wife and daughter it would be a fine end to an enjoyable Saturday evening spent over at his wife's brother's place a few miles away.

For other folks, the evening was spent in town. Teenage girls did babysitting duty in the scattered farmhouses; teenage boys planned either secret rendezvous with babysitters or a general rise in the level of hell via hotrods or maybe some leftover bottle rockets from the Fourth. But for most folks, Saturday night was for shopping for supplies, dinner at the cafe, a movie for the kids, adults meeting friends on the courthouse grounds downtown and talking of everything from Russians and Beatles to murders and snipers to soybeans and who would be Johnny's English teacher this year. At Prichard Auditorium, people were still hard at work setting up exhibits for the next day's 4-H Club annual fair, part of an entire week of activities collectively known as Horseplay Days. And out at the Country Club, a birthday party with some rare "imbibing" was moving along at fullbore. A fine time was had by all, but "it's already 'leven o'clock and it's high time we was in bed. Got church tomorrow, too!" The town square began to empty rapidly and Falls City, a small Nebraska town of 5,600, prepared to turn in for the night.

The farmer's old sedan rocked along the bouncy dirt road as it had so many times before, up and down over the gently rolling prairie hills. He noticed a line of storms stretching across the northern sky, slowly moving south-southeastward. The leading edge of the storm was like a shelf, with dark, roiling clouds piled on top. At the moment, the countryside around the car was still in the clear; the moon lit everything up so they could see everything for miles around. The farmer wondered if he even needed his headlights on, the moon was so bright. As he drove, a few low clouds came scudding in at right angles to the car, pushed along in front of the large oncoming cloud mass. The storm front itself was pulsing with explosions of lightning inside. The oblivious moon kept playing its hide-and-seek game with the smaller cloud puffs, unconcerned about the front's approach.

In the car, the family was getting close to home now, seven miles northeast of Falls City. Lightning bugs flashed and crickets and cicadas were louder than the car. The family could see the big mercury vapor light in their front yard now. As they watched the moon and kept an eye on the approaching front, they saw the red and white flashing strobes of an airplane coming up from the southeast. It was a jet, flying fairly low and imitating the moon by ducking in and out of the low clouds. The moon made the plane's white wings and T-shaped tail stand out, the effect enhanced by the row of lighted windows along the speeding jet's side. The good humor was still hanging about the car as the plane entered another cloud. In another instant, the entire night sky seemed to burst into flame—brighter and more sudden than even the lightning to the north. The laughter in the car stopped, choked off in constricted throats.

"My God, the world is on fire!" the farmer's wife screamed, finally recovering speech.

The farmer stopped the car and they sat, paralyzed and mesmerized by the sight of a ball of flames dropping almost straight down from the turbulent sky. A glimpse of a wing tip sticking out of the blazing comet as it fell told the farmer that the plane they had glimpsed seconds earlier was now in big trouble. Their mouths dropped open as a whirling mass of metal and flame came slowly spinning like a top through the air in front of them. The light was so intense it burned itself into their retinas and filled up their entire field of vision.

Just as the farmer thought the wreckage would hit their car or their house or both, it pancaked into a field up the road with an extremely loud bang, like a fist smacking onto a kitchen table. Shock waves rocked the car and wrenching metallic sounds assaulted their ears. The bang was followed by a massive flare of flames and debris shooting sixty feet into the air. He didn't remember hitting the brakes, but he must have; the car was idling on top of a hill looking down on the scene just a quarter-mile away. And as they watched, too stunned to move, the fireball began slowly collapsing in on itself. Small, intense fires now dotted what the farmer realized was his own soybean field, just across the road, yards from his own house.

Flashes of lightning from the oncoming storm and the waxing and waning moonlight illuminated the scene, but the mercury vapor yard light had gone out, its automatic light sensor telling it that dawn had come already. The ball of light still imprinted on their retinas, the three people in the car were roused by a loud clap of thunder. The storm now looked more menacing than before, as if its anger had been ignited by the hapless jet. The family could see that the gust front was rolling forward, an unstoppable force looking to do mischief to whatever it touched. The family also realized that the still-burning remains of the plane were close—way too close—to their home and outbuildings and livestock.

Antone Schawang awoke from his stunned state, hit the accelerator and sped forward, the car's rear-end sliding around on the loose gravel, his heart in his throat in anticipation of what he was about to find in his soybeans.

Bloomington, Minnesota
6:15 a.m., August 7, 1966
Sunday

He woke up early on that particular Sunday, shutting off the alarm clock before it could ring. Like so many other kids in the Twin Cities area, nine-year-old Tim Hilliker had stayed up later than usual the night before, thanks to summer vacation and a beautiful Saturday evening, but it didn't put a damper on his usual ritual. Sunday mornings for Tim meant waking up early to watch the *Gene Autry Show*. Syndicated reruns of the 1950s favorite played each week on a local Minneapolis television station, and Tim always set his alarm clock for 6:15 a.m. This would give him time to get downstairs, turn on the television to warm up and go find some breakfast. The show would start at 6:30 a.m. By that time the TV would be warm, and he would be eating to the sound of gunshots and galloping horses.

On this particular Sunday, the last moments of the "before" time, Tim jumped out of bed and headed out of the bedroom he shared with his younger brother, David, 8, who was still sound asleep. Tim threw himself down the stairs and jumped the last two steps into the living room of the Hilliker house at 10331 First Avenue in Bloomington, Minnesota, ten miles south of downtown Minneapolis. Tim hoped that his older brother and sister were, like David, still asleep and that his mother, Patricia, was drinking coffee and reading the Sunday *Star-Tribune* in the kitchen, giving Tim a rare moment alone with the TV.

Tim assumed his father was still sleeping and made a silent vow not to turn the volume too loud. James (Jim) Hilliker was a first officer for Braniff International Airways and had been on duty for a long flight segment the previous evening. His day was scheduled to end back home in Bloomington well after midnight. And when that kind of schedule was in the offing, Tim and his siblings knew to keep things quiet. At least for a while. But on this Sunday morning, when Tim hit the bottom of the staircase and started towards the TV, he realized something was very different. Instead of an empty room, he found his mother sitting in a chair in the dining room, looking into the living room and crying. His older brother, Dan, 15, and sister, Kathleen, 12, were not in the room.

He came to a halt in the middle of the living room and stared at his mother. He started to speak, but his uncle Bob Lindseth, his mother's brother, quickly took him aside and quietly broke the news. Some 450 miles to the southwest of their Bloomington home, while he slept peacefully, Tim's father and 41 others perished when their aircraft encountered a violent thunderstorm near a place called Falls City, Nebraska. This was the moment when nine-year-old Tim learned that his world had changed dramatically while he slept. This was the dividing point; everything that had happened in his short life up until that point was "before" and everything from that moment on was "after." Life would now be measured that way for Tim and his family: before and after Jim Hilliker's death.

Tim stood and stared at the floor, unable to comprehend what his uncle was saying. He thought that if only his legs would move, he would be able to run up to his father's bedroom and his father would be there, laughing and messing up his hair and telling him it was all a joke. His father was a renowned prankster. But Tim didn't move. Something inside him knew his uncle was telling the truth. His father was gone.

Part I. "I don't sleep well nights"

1

The Day the Movies Almost Died: Pan American World Airways Flight 115

I thought my jaws and shoulders were leaving my body.
—Passenger aboard Pan American World Airways Flight 115

Over the North Atlantic, 52.2 degrees north and 40.5 degrees west
10:00 p.m., February 3, 1959
Tuesday

The night of February 3, 1959, was a bad one for flying. The crew of an American Airlines 707 en route between New York and Los Angeles reported trouble with the aircraft's hydraulic system. Captain Harry Clark requested an emergency landing at Los Angeles International Airport, which deployed emergency equipment. Eventually, the crew determined that the fault was in the 707's instruments and they made a safe landing with no further delays or injuries.[1] But other flights on that night were far less fortunate. In Arkansas, a U.S. Air Force B47 bomber impacted nose down into rough terrain some 25 miles northeast of Little Rock Air Force Base. The three-man crew died.[2]

In an even worse accident in New York, a brand-new American Airlines Lockheed L-188 Electra landed short of the runway at LaGuardia Airport, crashing into the East River. The accident occurred just before midnight in heavy rain and fog. The Electra was just ten days old. Of the 73 passengers and crew members onboard American Flight 320, 65 died, including the captain and one of the stewardesses. An eight-year-old boy, Robert Sullivan, was the only survivor from his family. His father and two sisters died in the crash.

Robert and his mother swam from floating wreckage to a tugboat which was looking for survivors. Both were taken by an ambulance to Flushing Hospital, where his mother died three hours later.[3]

Civil Aeronautics Board investigators believed the crew became fixated on "particular aspects of the aircraft and its environment" and neglected instruments detailing attitude and height. This was complicated by the weather, the limited experience with the Electra by the crew, the use of the autopilot at too low an altitude, and sensory illusion due to limited lighting in the approach area.[4]

But even these tragedies, big as they were, had nowhere near the impact that a crash near Clear Lake, Iowa, had on American culture. In an accident still discussed today, three of American music's star performers died when the Beechcraft Bonanza they were flying to their next concert crashed in dark, snowy weather. Buddy Holly, 21, Ritchie Valens, 17, and J.P. "The Big Bopper" Richardson, 25, died along with the pilot, Roger Peterson, 21, of Clear Lake. The three performers were touring 24 cities in three weeks with Waylon Jennings, Tommy Allsup, Dion DiMucci, and Carl Bunch. The owner of the Surf Ballroom in Clear Lake, site of their last performance, had chartered the aircraft to save performers a lengthy bus ride to their next concert in Moorhead, Minnesota.

In a somewhat eerie similarity to the situation with American Airlines Flight 320 the same night, the pilot of the Bonanza became disoriented by the absence of lights in the area, as well as being unfamiliar with the attitude indicator. The aircraft hit the ground in a right-wing down attitude and tumbled for 570 feet. The pilot's body was still entangled in the wreckage, while Richardson was thrown into a neighboring cornfield and Valens and Holly were lying nearby.[5] In 1971, singer Don McLean wrote "American Pie" about the loss of Holly, Valens, and Richardson. In the song, McLean dubbed February 3, 1959, "the day the music died."[6]

But even with 65 dead in New York, three in Arkansas, four in Iowa, and the (fortunately short-lived) emergency over Los Angeles, the terror of the day was not entirely finished.

The previous day in Paris, American entertainer Gene Kelly and actress Susan Oliver were persuaded by a colleague to make a change in their airline reservations for the trip back to the United States. In the Pan American World Airlines office on the Avenue des Champs-Élysées, Kelly and Oliver were reconfirming their tickets on a February 3 Pan Am flight aboard a slow but steady propliner (most likely either a DC-7 or a Boeing Stratoliner). Writer-producer Harry Kurnitz was also in the office; he urged the other two Hollywood acquaintances to change their reservations to Flight 115, one of the new Boeing 707s which had been placed in service at Pan Am the previous October.

Kelly at first resisted, doubting that the new jet, which was extremely popular, would have available seats on such short notice. Pan Am's personnel found two open seats (nothing was said regarding whether the seats were actually open or if the internationally famous Gene Kelly, a Parisian favorite, could have whatever he asked for), and the seats were confirmed. Kelly was surprised and delighted. He had been visiting his daughter Kerry at her school in Switzerland over the Christmas and New Year holidays, but it was time to get back to work and anything that meant a shorter flight over the Atlantic was more than welcome.[7]

The next evening, Kelly, Oliver and Kurnitz, along with 116 other passengers and ten crew members, boarded Pan American World Airways Flight 115 for New York International Airport at Idlewild. This trip was operated on almost-brand new Boeing 707–121, N712PA, which had just 705 total flight hours on its airframe.[8]

The new jet was amazingly fast but not immune to laws of physics or aerodynamics. While it technically had the range to fly from Paris to New York nonstop, running into headwinds or the jet stream on the westbound side or being dispatched with heavier weight could require refueling stops. With thunderstorms possible on the route in early February, Pan Am 115 announced to its passengers that fuel stops at London Heathrow and Gander, Newfoundland, would be required.[9] The grumbling was minor; the jet would still beat pistons across the Atlantic to New York.

Two of Pan Am's veteran pilots were in the cockpit. Captain Waldo Lynch was pilot-in-command. In the right-hand seat was Captain Samuel Peters. Both were supervisory pilots; Peters was the captain in command of the airline's Pacific/Alaska division. Captain Lynch, 46, had type ratings in aircraft from the DC-3, -4, -6, and -7, as well as the 707. There were 11,185 hours in his logbook, 350 of them in the Boeing. Captain Peters, 49, had 269 hours in the 707 and 14,952 overall, in addition to the usual type ratings and certifications. George Sinski, 44, was the flight engineer, and John Laird, 41, was the navigator. Also in the cockpit was a company dispatcher named Mackey (no first name was given in the Civil Aeronautics Board [CAB] report), based at Idlewild. In the cabin, there were two pursers and four stewardesses looking after the passengers.[10]

After a normal departure from Paris and a normal refueling stop at London, the flight headed west. Kelly and the other passengers settled in for dinner, then drinks, followed by conversation and napping. The aircraft was speeding along on autopilot at 35,000 feet in smooth air at 10:05 p.m. GMT (Greenwich Mean Time), all systems normal, when Captain Lynch decided to visit the main cabin. Peters remained in the right-hand seat, with the other three in their assigned crew positions.[11]

What happened next was described in the usual clinical way by CAB investigators. After Lynch exited the cockpit,

> the autopilot disengaged and the aircraft smoothly and slowly entered a steep descending spiral. The copilot was not properly monitoring the aircraft's instruments or the progress of the flight and was unaware of the actions of the aircraft until considerable speed had been gained and altitude lost.
>
> During the rapid descent the copilot was unable to effect recovery. When the captain became aware of the unusual altitude of the aircraft he returned to the cockpit and with the aid of the other crew members was finally able to regain control of the aircraft. Recovery was made at an altitude of approximately 6,000 feet.[12]

The CAB report may adequately cover the event and be written in language appropriate to its purpose. But it is, shall we say, a little understated.

Gene Kelly granted an interview a few days later to Peer J. Oppenheimer, *Family Weekly*'s Hollywood correspondent (the magazine was a newspaper supplement for Sunday morning editions nationwide, much like the *USA Weekend* or *Parade* inserts of today). Kelly's story was told in the first person in the April 5, 1959, edition of the magazine, in an article entitled "Gene Kelly's Brush with Death."[13]

It's interesting to compare the CAB report with the Hollywood star's account. Where investigators could be clinical and detached, Kelly, even two months later, was not so sanguine. Kelly wrote he was "slouched into seat 1D" in the front row behind the forward lounge. He was "awed" by the 707, but was comfortable, even though it was his first trip on a jetliner, as well as the highest in altitude he'd ever flown. The takeoff from Paris was "thrilling" and the food and drinks "delectable." Even with the announcement that there would be weather-induced delays and stops in London and Gander, Kelly's impatience turned to calm and complacency. Harry Kurnitz was two seats behind; there was no mention of where Oliver was sitting. Interestingly, Kelly admitted he was not wearing a seat belt, "naturally."

Kelly had conversations with the flight crew and investigators later, and they filled him in on what had actually happened. It suddenly felt like the jet was headed straight down into the ocean. He reported that he, like everyone else, was pushed down into his seat and experienced "excruciating pain" in his ears (it would take more than two days for his ears to stop hurting). The sensation was akin, he said, to being thrown to the roof of an elevator and being held there after the cables snapped, falling downwards at high speed. He noticed that some crew members who were going about their usual duties were thrown to the floor of the cabin and pinned down.

Kelly was convinced that a crash into the Atlantic would occur within seconds. As with many others aboard, he went through a variety of emotions;

one of the first was anger at himself for switching his reservation to Flight 115 just to ride "Jet Clipper Washington." He believed that everyone on board would disintegrate in the water and that they, and the cause of the crash, would never be found. He repeated the usual question/exclamation: "Why? Why!"

As the aircraft dove faster, pressure built on everyone's eardrums. Kelly said his felt as if they were being pierced with "hot needles." His hands were immobilized at his sides and he could not even buckle his seat belt. As the seconds ticked by, Kelly began to think about longer-term issues; with the pain in his ears, would his hearing be damaged so that he would never hear again? Would there be at the least hearing difficulty all the remainder of his life? He also began to think his seat belt should be secured; in a crash, a seat belt might keep him from being thrown about the cabin and killed. The effort seemed to last hours. Kelly reported that there were no screams or other loud noises from the other passengers and crew members, no apparent hysterical reactions.[14]

Later in New York, Kelly, Kurnitz, and Oliver told syndicated columnist Leonard Lyons what they were thinking during the sudden dive. Kelly told Lyons he condemned Kurnitz for convincing him to switch, and also wondered if his insurance was current. Kurnitz also supposedly was concentrated on insurance, claiming he thought he would love to settle with his agent for "two cents on the dollar." And Oliver made a joke out of the whole thing: She said she wasn't ready to die because she wanted to have at least the chance at better billing with Gene Kelly. "The headlines will just say 'Gene Kelly Killed,'" she quipped.[15] But the jokes would come later.

Meanwhile, on the floor in front of him, Kelly saw Captain Lynch get thrown to the floor. Only by an extreme effort was Lynch able to crawl on his stomach toward the cockpit door.[16] CAB investigators told what was taking place on the other side. Captain Peters turned to a chart required by Pan American during the long crossings. Described as a "how goes it" curve, the report had to be filled out at certain times during these long trips and "required computations of time, distance, cruise speed, fuel consumption, some of which are computed by the navigator." This was entered into a report on a clipboard in Peters' lap. He was also wearing a headset covering both ears so he could hear the 10:05 p.m. GMT weather report from Gander. Concentrating on all these things (with the flight engineer and navigator concentrating on their own work), Peters didn't have his eyes on the dashboard instrumentation.[17]

The first time Peters realized something was not normal was when he felt a buffeting in the aircraft frame, "followed by a feeling that positive acceleration forces were building up rapidly. The buffeting increased in intensity and his instrument panel lights went out."[18] Peters looked over to his left at the captain's instrumentation, which retained its lighting. He "saw that the captain's artificial horizon had tumbled and consequently was of no use to him. He then glanced up and saw the stars moving rapidly counterclockwise, indicating that the aircraft was in a nose-down right spiral about to roll over on its back."[19]

It was at this moment that Peters "grabbed the control wheel, pushed the autopilot release button, and attempted to stop the roll by applying left aileron and rudder."[20] It was at this point in the dive, however, that G forces had built up so high that everyone was pinned down and immobilized. The cockpit environment had come alive with "various system-warning and fire-warning lights ... being activated intermittently and the Mach warning bell" ringing.[21] It was then that Captain Lynch managed to finish his crawl back into the cockpit. "As he passed the flight engineer, Mr. Sinski reminded him that the power was still at cruise thrust. The captain pulled the power levers to idle position and pulled himself into his seat."[22] The seats in most airliner cockpits are on rails; they get moved fully to the rear to allow the pilots to easily get in and out of them. When in place, the pilots' seats are moved full forward and rudder pedals in the floor are moved towards the pilots.

Just before G forces made it impossible, Mr. Mackey the dispatcher and Mr. Laird, the navigator, switched seats, so that Laird could help Peters. Once back in his seat, Lynch asked the navigator, now directly behind him, to hold him in his seat. None of this was easy. "Everyone in the cockpit was seriously affected by the G forces which made it difficult or impossible to move properly their heads, hands, or feet. Captain Lynch said that his head was bent over and his feet seemed pinned to the floor,"[23] investigators later wrote.

Lynch was able to get a "quick glance at his instruments." They "showed the airspeed needle in the vacant area to the right near the zero mark, and the altimeter passing through 17,000 feet with the needle turning at a terrific rate. He could not see the Mach meter because it was hidden by the control wheel and he could not lift his head. The artificial horizon was of no use to him because it had tumbled, and the turn and bank indicator was full to the right with the ball positioned slightly to the left of center." His instrumentation was either spinning crazily or completely useless. Lynch quickly glanced at Peters and, seeing him struggling with the controls, shouted, "I have command,"[24] in accordance with established procedures. The situation was not good: "The stabilizer was in the full nosedown position and [the Captain's] electric trim button failed to function. Visual reference was impossible because they were in a cloud."[25]

"Laird somehow managed to fasten the captain's safety belt and while this was being done Captain Lynch rolled the wings level and the G forces were relieved." They could now move and breathe easier. "The flight engineer, now able to move, immediately pulled the circuit breaker which deactivated the stabilizer system and then straddled the console and began rolling both stabilizer wheels toward the up position by hand."[26] The aircraft began to come out of its dive. Captain Lynch pulled the yoke back with a steady pull as they passed 8,000 feet, bringing up the nose further. The worry was that aircraft pieces, or the engines or the wings, would come off during the attempt to arrest the dive. But the aircraft, while heavily damaged, remained mostly intact and flyable. "At 6,000 feet there was a terrific violent pounding or buffeting which lasted a couple of seconds and then the aircraft ceased to descend and began a fairly steep climb. At 9,000 feet the wings were level and the aircraft was in a moderate climb."[27]

With the aircraft no longer in apparent danger of hitting the Atlantic nose first, Lynch "asked Flight Engineer Sinski to roll the stabilizer a bit forward and with the aircraft responding reasonably well to control demands he realized he had once again regained positive control. He then moved the horizon switch to the No. 2 position, selecting the No. 2 vertical gyro, and his horizon registered normal pitch movements, but depicted a steep bank." His aircraft was back under control, but his instruments were still a scrambled mess. After determining their position, the crew "immediately advised Gander … of the difficulty and a cruise altitude of 31,000 feet was obtained for the remainder of the trip."[28]

Back in seat 1D, with his ears still in pain, Gene Kelly was breathing again. He was able to hear Captain Lynch announce that the flight would be "okay," but that the crew was "too busy up front to go into details."[29] The cabin attendants began instructions on how to survive and evacuate the 707 in an ocean ditching. Passengers donned life jackets, the crew moved about with calm efficiency, and life rafts were placed at exit doors ready for deployment. Kelly and the other men removed ties and shoes and prepared to assume the "brace" position.

Kelly's thoughts then returned to regrets. He "scolded" himself and criticized himself for eating and drinking so much that his entire head was fuzzy. He was afraid he would be unable to do anything for himself, let alone anyone else who needed help. He also thought more about his insurance premiums, hoping that his mother would be provided for if he didn't come back. He also remembered a previous in-flight emergency, this time on a trip to St. Louis a few years previously. The aircraft's landing gear wouldn't properly drop down, so the crew circled the city for an hour. Emergency equipment was deployed for the landing. Kelly remembered another passenger on that flight referring to the assembled ambulances as "blood wagons." But he remembered how that aircraft landed safely without injury or incident. *So why not this one?* he asked himself. A few of the other passengers were given oxygen due to the sudden changes in pressurization during the dive and recovery. Kelly took a few seconds of oxygen in the hopes of easing whatever pressure was still bothering his ears. He reported that it did not work.

Captain Lynch announced continuation to Gander, this time at 31,000 feet. For over an hour and a half, Pan Am 115 flew on. The landing at Gander was uneventful. Kelly's earache and fear of the new jets subsided within days. Although the autopilot needed work, he said he was convinced that new jets were well built. He would fly them often. Once on the ground in Gander, the passengers were made comfortable at the airport and any in need of medical attention were given care while everyone waited for Pan American to send a replacement aircraft. Investigators reported that the incident aircraft "was carefully examined and it was determined that although it had sustained extensive structural damage it could, with minor repairs, be flown safely to the Boeing plant at Seattle, Washington, for final repair."[30]

Once the replacement 707 arrived at Gander, "all passengers and crew continued the flight to its planned destination."[31] But the incident aircraft stayed in Gander while officials from all interested parties gave it a thorough going-over. Said the CAB:

> The damage consisted mainly of buckles in the lower surface skin of the right and left horizontal stabilizers and buckles in the center section web and upper surface doubler, and both wing panels were damaged including shear wrinkles in the rear spar webs and damage to the outboard ailerons and aileron control rods. The wing-to-fuselage fairings were damaged and a three-foot section of the right fairing separated in flight. Both wing panels suffered a small amount of permanent set. All four wing-to-strut fairing sections of the engine nacelle struts were buckled. Nos. 2 and 3 nacelle shear bolts partially failed in shear and the fitting holes of all front spar-to-wing bushings were elongated.[32]

Basically, the wings were flexed upward some 17 feet with such force that they remained flexed even after the aircraft was back on the ground. Damage associated with this was found all over the structure, as described. Later estimates from Boeing experts calculated that the aircraft and its occupants had experienced G-forces of 6.7 times the force of gravity during the pull up out of the dive and that the aircraft had exceeded Mach .99 during the dive itself—far above what it was designed to do. The 707 was a tough bird.

As for why Flight 115 dropped out of the sky in the first place, "certain systems and components of the aircraft were definitely suspect" by the CAB, including "the autopilot, Mach trim systems and their warning systems." "Comprehensive tests of these systems were made at the Boeing Airplane Company plant near Seattle, Washington, under the direction of a CAB investigator."[33] First to be examined was the PB-20D autopilot, which provided "sensitive automatic, coordinated control of the aircraft." This examination turned up an interesting finding:

> On several previous flights of B-707 aircraft there have been disengagements of the autopilot without the existence of a mechanical failure and after which the system functioned normally when the autopilot was reengaged. One such disen-

gagement occurred on the immediately preceding flight of the subject aircraft [N712PA the day before it was used on Flight 115]. On this occasion the warning light failed to come on and the crew became aware of the disengagement by observing a 20-degree right-wing low attitude on the horizon indicator. Before recovery was effected approximately 600 feet in altitude was lost.

The warning light was checked by the crew immediately after the occurrence and it did not light. Functional tests subsequent showed it operating with two exceptions:

1. In several instances the autopilot disengage warning light did not function properly after disengagement of the autopilot.
2. The pitch trim potentiometer did not recenter after autopilot disengagement. The mechanical centering of this potentiometer is necessary for the autopilot upon reengagement to have available full nose up or nose down trim.[34]

However, even with a suspect autopilot, it is the duty of the flight crew to make sure the aircraft is being flown properly, even if by machine. The board ruled that the probable cause in this case was "the inattention of the copilot to the progress of the flight, during the absence of the captain from the cockpit, following the involuntary disengagement of the autopilot. Contributing factors were the autopilot disengage warning light in the dim position and the Mach trim switch in the 'off' position."[35]

Following the investigative process and finding of probable cause, the FAA, "citing failure to comply with Part 41.62 (Pilots at Controls) of the Civil Air Regulations, levied a civil penalty against the captain. The copilot received a six-months' suspension of his ATR [Airline Transport Rating]."[36]

The three passengers from Hollywood were happy to be on the ground back in New York and California. Already successful, they would all three go on to further success. Gene Kelly (1912–1996), was the quintessential American entertainer, with a career spanning over 50 years as an actor, dancer, singer, director, producer and choreographer and a list of credits far too long to list here. He is especially known for his biggest successes, *An American in Paris* and *Singin' in the Rain*. A year after his Pan Am 115 experience, he returned to France and became the first American choreographer to create a modern ballet for the Opéra-Comique. Called *Pas de Deux*, it incorporated George Gershwin's Concerto in F. For this, Kelly was made a chevalier of the French Légion d'Honneur.[37]

Harry Kurnitz (1908–1968) wrote extensively for movies and television, including *Shadow of the Thin Man* (1941), *The Thin Man Goes Home* (1945), *The Inspector General* (1949), *Witness for the Prosecution* (1957), and *A Shot in the Dark* (1964, the second in the Peter Sellers *Pink Panther* series).[38]

But it was perhaps Susan Oliver (1932–1990) who was impacted most by her experience on Pan Am 115. Oliver worked hard in the entertainment industry over the next three decades. She had extensive television credits, including directing an episode of *M*A*S*H* in 1982 and one of *Trapper John, M.D.*, in 1983, and appearing in the original pilot episode of *Star Trek*. Her acting career began in 1956 and included appearances in episodes of such TV icons of the era as *Father Knows Best*, *Bonanza*, *Playhouse 90*, *Twilight Zone*, *The Adventures of Ozzie and Harriet*, *Rawhide*, *Wagon Train*, *Route 66*, *Dr. Kildare*, *77 Sunset Strip*, *Gomer Pyle USMC*, and *Peyton Place*.[39] She also had a memorable turn as a woman prisoner lodged by the state police overnight in the Mayberry jailhouse on *The Andy Griffith Show*, during which she hoodwinks Andy, Barney and Aunt Bee and almost makes her escape.

In film, Oliver worked alongside Elizabeth Taylor in *Butterfield 8*, Joan Crawford in *The Caretakers*, Sal Mineo in *The Gene Krupa Story*, and Jerry Lewis in *The Disorderly Orderly*. In her book, *Odyssey: A Daring Transatlantic Journey*, Oliver described how her experience onboard Pan Am Flight 115 resulted in a fear of flying she overcame largely through hypnosis.[40] She returned to the skies with a vengeance in the '60s. She obtained a pilot's license, won the 1970 Women's Powder Puff Derby air race, and flew solo across the Atlantic (New York to Denmark) in a single-engine prop plane. Ms. Oliver worked in television until 1988. She died of lung cancer in 1990.

With all the death and carnage across the country and out over the North Atlantic on February 3, 1959, it's amazing "The Day the Music Died" was not also "The Night the Movies Died." As it was, that was a very near thing. Other jet upsets (regardless of the precipitating event) in the early "Jet Age" days would not end nearly so well.

Pan American World Airways Flight 115

Summary
Pan American World Airways, Inc.
Boeing 707–121, N712PA
52.2° north and 40.5° west, over the North Atlantic between London, England, and Gander, Newfoundland, Canada
Tuesday, February 3, 1959
Fatalities: None; a few very minor injuries

Report
Civil Aeronautics Board
Aircraft Accident Report
File No. 1–0006
Adopted October 28, 1959; released November 3, 1959

2

Melting Into a Threatening Sky: Northwest Orient Airlines Flight 705

We can see it out ahead ... it looks pretty bad.
—Flight crew, Northwest Orient Airlines Flight 705

---

37 miles west-southwest of Miami International Airport, Florida
1:50 p.m,. February 12, 1963
Tuesday

The advent of the Jet Age brought with it high expectations in the commercial aviation world. Not only would it be the fastest way to get from point to point, it would also be the smoothest and most sophisticated and most comfortable. The "smooth air" promise was one of the biggest selling points for passengers. The Boeing Airplane Company produced a promotional video at the end of 1959 highlighting the first year of service of the company's first jetliner. Titled *707: Year One*, the film emphasized the jetliner's speed, structural strength and record-setting capabilities. The film's narrator states: "The Mississippi River of Huck Finn is just two-and-a-half hours from New York!" and "Now the businessman can be a bi-coastal commuter! Breakfast in San Francisco, Lunch in New York, and home again, all in the same day!"[1]

But, while speed was indeed important to passengers, especially in the business community, the promise of flights that would no longer feel like riding the Coney Island Cyclone ("Riding in smooth air, above the clouds, in sunshine all the way,"[2] was Boeing's promise). was particularly compelling to millions of passengers who had experienced almost four decades of bouncing around the sky in propeller-driven aircraft which could not get above the worst of inclement weather. And the promise of smooth air was not just a marketing fantasy. Experts had the same expectations. As the Jet Age proceeded to get underway in the United States in 1958, researchers came together to discuss the "smooth air" promise. The National Air and Space Administration (NASA) convened the NASA "Conference on Some Problems Related to Aircraft Operations." Roy Steiner and Martin R. Copp, researchers at the Langley Research Center (later a part of NASA), presented a paper at the conference titled, "A Review of Atmospheric Turbulence and Its Significance to Jet-Transport Operations."[3] The paper summed up what was known during the months before Pan American World Airways launched the first U.S. commercial jet service:

> Gust and operating data have been collected from airline operations for a number of years. The results obtained from these data-collection programs, together with data from other sources, have served to develop a description of the turbulence in the atmosphere. The purpose of this paper is, first, to review this basic information on the turbulence, especially in regard to its intensity and frequency of occurrence at different altitudes, and second, to examine the significance of the turbulence to jet-transport operations.[4]

The report went on: "It is expected that jet transports, at their higher cruising altitudes, will fly above or around most of the severe cloud turbulence and, because of their longer flight lengths, will climb and descend less frequently through these lower altitude regions. It appears, therefore, that jet transports will experience less storm turbulence than current transports, but the relative amount of clear-air turbulence is not apparent because of jet-stream turbulence which may be encountered at the high cruising altitudes."[5] In a section regarding "Turbulence Avoidance," the report noted the following:

> These acceleration estimates for the jet transport would be modified if the operating airspeeds differed from the assumed speeds. Storm-avoidance procedures, either visually or by use of radar, may also materially affect the accelerations. There are no operational data available from radar-equipped jet transports, but some indication might be obtained from a review of past radar results. The results from one investigation on the use of airborne radar for storm avoidance on a low-altitude transport operation ... are shown [here]....
>
> [The figure] shows the gust accelerations for piston-engine transports flown by the same operator. These two curves represent the data before and after radar was installed in the airplanes. The average cruising altitude and flight length were similar for the two operations, as indicated in the figure. The curves indicate that, for the larger gust accelerations, the values were about 25 percent lower when radar was used for storm avoidance. No reduction was obtained for the small accelerations. These same reductions may not necessarily apply to

jet transports, but there is some indication that the large loads may be reduced somewhat for jet operations by radar storm-avoidance procedures.[6]

This assumption, that radar would help jet transports avoid turbulence, was based on the development of airborne radar since World War II. Radar turned to peacetime use after the war and the technology was gradually added to commercial aircraft during the 1950s. In the period the report is covering, some airlines even put "radar-equipped" or similar signs on the front of their aircraft to advertise to passengers that their flight would seek out the smoothest and safest route to their destination. In other words, gone would be the bad old days of stewardesses having to run up to the cockpit in order to get a breath of fresh air and respite from the, shall we say, "intestinal upsets" in the passenger cabin caused by roller coaster rides in bumpy weather. Not only would technology be harnessed by the airlines to enhance safety, but it would also be used to make the "air sickness bag" obsolete.

More seriously, however, the report put forward an interesting prediction in its summary: "An application of these data to prospective jet operations indicates that: (1) the gust experience may be reduced significantly relative to current operations, and (2) a somewhat smaller reduction might be expected in the acceleration experience. These reductions become more significant as the flight length is increased."[7] It was a reasonable, logical and historically based conclusion. But was it correct?

Northwest Orient Airline's timetable for February of 1963 features a photo of legendary Hollywood comedian Buster Keaton decked out for the beach in a striped bathing costume from the 1920s and holding a patched rubber inner tube and a Northwest flight bag, with his signature pork-pie hat on his head and a frowning face that shows his disgust with brutal winter. He's on his way to a sunny beach, thanks to Northwest. The text reads, "Get in the Sun.... Take the fun way to Florida. Enjoy Regal Imperial Service and choose from 10 Fan Jets daily between the midwest and Florida!"[8] The airline's flights between the two areas are then summarized in their own special "NWA Florida Quick Reference Schedules" next to Keaton's picture.

At 12:40 p.m. on February 12, 1963, Northwest's Boeing 720B, N724US, operating as Flight 700, arrived at Miami International Airport from Chicago, leaving behind a 22-degree wintry day with 10-mile-an-hour winds.[9] Billed as a "720B Fan-Jet" flight, the trip south featured "Regal Imperial Service," as advertised in the timetable. The service provided "at no extra fare, specially selected menus and passenger service features, including complimentary Florida beach bags." Champagne, imported wine and cocktails were offered to first class passengers; tourist passengers could purchase cocktails.[10] All of this was designed for a traveler like Keaton who needed to soak up the warmth of south Florida's beaches and sunshine. Weather in south Florida was indeed warm (79 degrees at noon) and spring-like. In fact, the U.S. Weather Bureau said Miami-area weather was "characterized by a pre-frontal squall line approximately 250 miles in length, oriented on a northeast-southwest line immediately northwest of the city." The bureau's radar observation at 1:44 p.m. showed a "broken area of thunderstorms associated with this line, with cells two to twenty miles in diameter, and tops of detectable moisture at 30,000 feet. The line was moving southeast at eight knots, and moderate rain showers were occurring at the [Miami] station."[11]

With such a wide front spread across the state, air traffic was affected. The inbound Flight 700 was thirty minutes late, so ground crews swarmed the almost-brand new jet and quickly prepared it for its next trip, which was scheduled to depart less than an hour later as Flight 705. The airliner would return to Chicago, then proceed westward to Spokane and Seattle, Washington. Just 35 passengers were scheduled for the northbound trip[12] and many of them were returning home after their own warm Winter-in-Florida vacations and leaving behind the chance to use their own bathing suits and inner tubes. As if to underline how the outbound trip was a return to the Midwest's winter reality, no "Royal Imperial Service" was offered for Flight 705. There was also no comical picture of Buster Keaton heading to Chicago in the timetable.[13]

Northwest's ship number 724 for the Miami service that day was officially a Boeing 720–051B, which first flew on July 14, 1961, and was delivered to Northwest on July 26.[14] It had been in service for only six months when it was involved in a landing accident at Fort Lauderdale, Florida. The Civil Aeronautics Board reported that the aircraft, while operating a flight from Miami to Seattle with stops at Fort Lauderdale on January 26, 1962, landed short of a runway: "Structural failure occurred when the right main landing gear separated, with resultant damage to the adjacent wing, flap, and fuselage areas, and the No. 3 and No. 4 engine nacelles." In its official probable cause statement, the agency blamed the pilot-in-command, who "misjudged [the] distance and undershot [the runway] during landing. [The right main landing gear] struck a two-foot rise on the blast [concrete area] 384 feet short [of the runway threshold]. The official report lists the Fort Lauderdale trip as having seven crew members and just two passengers aboard, none of whom were injured.[15] Ship 724 was repaired and returned to service. It sustained a bird strike on the right wing leading edge at some point in the next year, and that damage was also repaired. The CAB reported these two incidents as "the only occurrences of significant structural damage to the aircraft [before its final crash]. The maintenance records reflect compliance with FAA Standards of airworthiness."[16] At the time of its last trip, the aircraft had a total flight time of 4,684 hours. Its four Pratt and Whitney JT3D-1 turbojet engines had between 2,200 and 3,600 hours of flight time.[17]

Upon arrival in Miami February 12, 1963, the captain of inbound Flight 700 flagged a minor maintenance item for mechanics to look at during the turnaround. N724US's "outflow valves [were] a little sticky [which] merely made it a little difficult to maintain the [cabin] pressurization in a smooth manner." Mechanics cleaned the valves, remedying the problem. A leaking rivet at the number 4 reserve fuel tank was plugged when it was noticed by the mechanic."[18] The aircraft was ready for its 35 passengers and seven fresh crew members.

While the aircraft was being prepared for its return to the northern winter lands, Flight 705's cockpit crew were in the airline's operations center. Flight 705 was to be helmed by Captain Roy Almquist, an experienced, accomplished pilot with just over 17,835 hours of flying single and multiengine aircraft. Almquist was a 47-year-old born in Minneapolis. Single and a graduate of Roosevelt High School, he had joined Northwest in 1942. He lived in Rosemont, Minnesota, not far from Minneapolis International Airport. Almquist was certified for flying Douglas DC-3s, DC-4s, DC-6s, and DC-7s; Lockheed L-188 Electras; and the Boeing 720B. Since the latter was so new (the first example of the type first flew on November 23, 1959), Almquist had only 150 hours of flying time in it. He passed his latest flight proficiency check on November 13, 1962, and his FAA first-class medical certification on November 21.[19]

That last initial flight proficiency check began on November 9. For three hours and 50 minutes examiners had Almquist perform a series of prescribed maneuvers in the 720B, but the flight check was terminated early because of "mechanical difficulties" with the aircraft. Almquist's check continued with a two-hour flight on November 12, and an hour-and-24-minute flight on November 13. A Federal Aviation Agency (called after 1966, Federal Aviation Administration, FAA) air carrier operations inspector rated Almquist on 22 different items, with a possible grade/range from 1 (well above average) to 5 (unsatisfactory). The lowest passing grade was 4. That grade of 4 was given by the FAA inspector for Almquist's performance on nine of the 22 items checked, which included how he recovered from situations involving Dutch rolls ("complex oscillating motions of an aircraft involving rolling, yawing, and sideslipping"), jammed stabilizers, electrical emergencies and engine fires.[20]

Flight 705's first officer was 38-year-old Robert J. Feller of Minneapolis. Feller held a valid airline transport pilot certificate (No. 500934) and was certified to fly DC-4, DC-6, DC-7, and 720B aircraft. He was also an accomplished pilot, with 11,799 hours of flight time, 1,093 of it in the 720B. Feller had passed his last flight proficiency check on July 8, 1962; his FAA first-class medical certificate was dated October 4, 1962.[21]

The Boeing 720B, like many of its fellow early jetliners, had a three-man cockpit. In addition to the captain and first officer, there was a flight engineer, sometimes called a second officer, responsible for monitoring and adjusting the aircraft's systems and performance. Flight 705's second officer was 29-year-old Allen R. Friesen of Hopkins, Minnesota. Friesen held both a valid airline transport pilot certificate (No. 1246257) and a flight engineer certificate (No. 1,492,889). All of his total pilot flight time of 4,852 hours and his 523 hours as a second officer were accumulated on the Boeing 720B. Friesen had passed his last flight proficiency check on May 8, 1962; his FAA first-class medical certificate was dated April 18, 1962.[22]

While the airline may not have offered "Royal Imperial Service" on its northbound flights from Florida, Flight 705 still offered a high-quality passenger experience, with five flight attendants (still called "stewardesses" in 1963) looking after the passengers, four of the stewardesses veterans and the fifth with just two months of service. All five lived in Minneapolis:

- Virginia Lee Younkin, 25, hired June 16, 1958, and qualified for the Boeing 720B June 23, 1961;
- Myrna E. Ewert, age 28, hired April 24, 1959, and qualified for the 720B June 19, 1961;
- Wendy F. Engebretson, age 21, hired September 29, 1961, and qualified for the 720B September 26, 1961;
- Connie Rae Blank, age 21, hired April 28, 1962, and qualified for the 720B on April 21, 1962;
- Mary S. Sandell, age 20, hired December 22, 1962, and qualified for the 720B December 19, 1962.[23]

Captain Almquist gathered as much weather information as possible prior to going out to the tarmac. He discussed the situation with the flight crew who had just landed on Flight 700. The captain of that flight told Almquist "that the weather extended from LaBelle, approximately 70 miles northwest of Miami, to the Miami VORTAC. The tops of the clouds were estimated to be at 27,000 to 30,000 feet."[24] (A VORTAC is an aeronautical navigation aid consisting of a "very high frequency omnidirectional radio range" (VOR) and an "ultra high frequency tactical air navigation (TACAN) radio range"; they provide course and distance information to pilots.) Basically, the front ran diagonally from northwest to southeast from near Fort Myers down to Miami. When appearing before the CAB several months later, the captain of Flight 700 also testified that "I simply explained to him [Almquist] the weather as I saw it approaching the front, and I explained to him how we had been cleared over the weather and made our letdown to the east side of the frontal area."[25]

In addition to discussing the weather situation with the arriving crew and Northwest's dispatchers, Almquist had a copy of an alert from the U.S. Weather Bureau known as a SIGMET (an abbreviation for "significant meteorological information"), which is "a message designed primarily for aircraft in flight, warning of weather conditions potentially hazardous to transport category and other aircraft." Specifically, the weather bureau issued SIGMET No. 3 of February 12, which was valid from 9:00 a.m. to 1:00 p.m. for the Miami area. It forecast "moderate to severe turbulence in thunderstorms, with a chance of extreme turbulence in heavier thunderstorms."[26] Northwest's operations agent at Miami called SIGMET 3 to the attention of 705's crew and attached a copy of it to the flight dispatch papers.

Not long afterwards, the weather bureau issued SIGMET 4 for the day, covering the time between 1:00 and 5:00 p.m. Flight 705's crew was not given the updated information sheet; it arrived in the Miami airport dispatch office at 1:15 p.m. after they had left the operations office to board the aircraft. SIGMET 4 predicted "moderate to severe turbulence, but deleted the reference to extreme turbulence indicated in SIGMET No. 3." CAB investigators noted that, "since the dispatcher for this flight is stationed in Minneapolis, the physical limitations involved made it difficult to apprise the crew of this latest advisory prior to their taxi time of [1:25 p.m.]."[27]

At the time, the *U.S. Weather Bureau Manual* categorized turbulence as being of three types: moderate, severe and extreme. It defined moderate turbulence as rough air causing seat belts to be required and "unsecured objects [to] move about." Severe turbulence would result in aircraft being "out of control momentarily" resulting in occupants being "thrown violently against the seatbelt and back into seat." The last category would be very rare; extreme turbulence would result in aircraft being "violently tossed about [and] practically impossible to control." Damage to aircraft structures could result from an encounter with this last category.[28]

Northwest issued its own route weather forecast for points south of Chicago, effective at 1:00 p.m. information that was also reportedly seen by Almquist and his crew. The airline's forecasters reported the cold front at Fort Myers, "moving eastward at 20 knots, with a line of thunderstorms 100 miles east of the front." The forecast noted that clouds encountered on routes south of Macon, Georgia, would be around 25,000 feet high, with "a few thunderstorms to 40,000 feet in the Miami area." Northwest's forecast made no specific reference to turbulence. But, as the CAB noted, the "company meteorologist who prepared the route forecast for Flight 705 stated that turbulence was indicated in his forecast by the presence of convective clouds." And, in fact, Northwest's flight operations manual stated that "if cumulus clouds are forecast to exceed 10,000 feet severe turbulence may be expected."[29] Northwest's definition of "severe turbulence" was in line with the U.S. Weather Bureau definition.

Once aboard the aircraft, 705's flight crew performed the usual preflight checks and readied for departure. They radioed the ground controller to ask about the departure routes being utilized, and he replied that most flights were departing "either through a southwest climb or a southeast climb and then back over the top of it [the line of bad weather]."[30] They would need to divert east or west to either of Florida's coasts before heading back to the north.

Scheduled to leave Miami at 1:25 p.m. Eastern time, with arrival in Chicago 2 hours and 40 minutes later (at 3:05 p.m. Central time), Flight 705 was only slightly late on February 12. The Miami ground crew made up for the inbound flight's weather delay, and Almquist was able to move the 720B away from the gate just ten minutes late. He advanced the throttles and the big jet roared down the runway. According to its flight data recorder, N724US was airborne at 1:35:22, making an initial left turn after departure from Miami International's runway 27L.[31] As Frank Tinker would put it in *Popular Mechanics* later that summer, Flight 705 "took off that afternoon and melted into a threatening sky."[32]

Miami departure control gave the flight a circuitous routing using radar vectors, directing 705 on a similar path as an earlier departing flight. At 5,000 feet on a heading of 300 degrees (pointing to the west of northwest), with Captain Almquist flying and First Officer Feller handling communications duties, 705 requested clearance to climb to a higher altitude and started a conversation with a radar departure controller about storms ahead. The controller coordinated climb clearance with the Miami Air Route Traffic Control Center (ARTCC). During the conversation, Feller told the controller, "Ahh, we're in the clear now. We can see it out ahead ... looks pretty bad."[33]

Eight minutes after departure, the controller cleared 705 to climb to flight level 250 (25,000 feet). Feller acknowledged the clearance and said, "Okay, ahhh, we'll make a left turn about thirty degrees here and climb." The new heading would be 270 degrees (to the west), which would take them "out in the open again."[34] Clearance for the move was granted. During the climb, the controller and Feller discussed the severity of the turbulence being experienced, which Feller described as "moderate to heavy." He also told the controller, "OK, you better run the rest of them off the other way then."[35]

The advice coming from 705's crew was not idle chatter. Conversations such as this one are an important part of the way information about flying conditions, especially turbulence, is shared among flight crews. Called a PIREP (pilot report), the information is radioed to the nearest ground station, encoded and relayed to other weather stations and air traffic. In the case of Flight 705, the PIREP gave the flight's location, heading, altitude and speed as well as the turbulence information, all of which was then relayed to other flights in the area, whether on the ground or already airborne, so that routings could be devised to stay away from the worst areas.³⁶

Two minutes later, at 1:45 p.m. the Miami radar departure controller directed the flight to contact the ARTCC on an assigned radio frequency. At first, the flight did not respond. Miami departure control then gave them a secondary frequency for ARTCC and also instructed the flight to turn to a heading of 360 degrees (to the north). This time, Feller responded affirmatively and switched frequencies, contacting the en route center.

When Miami ARTCC requested its position and altitude, Feller replied, "We're just out of seventeen five (17,500 feet) and standing by on the DME one."³⁷ It was 1:48 p.m. just 13 minutes after takeoff. This would prove to be the final conversation received from Northwest 705.

3

"Unusual conditions and circumstances": The Investigation of Northwest Orient Flight 705

An orange ball of flame ... dropped straight down, becoming a streak.
—Eyewitnesses, Little Banana Patch, Florida Everglades

Multiple Locations
1963–65

At around the same time as 705's last communication, ground controllers were receiving PIREP-type communications from personnel on other flights, four of whom were interviewed later by CAB investigators. The first was from the crew of a large jetliner flying at 7,000 feet and approaching Miami from the west. This crew reported that they were "in and out of broken clouds and light rain showers with light turbulence. Darker, heavy shower activity was observed to the (south) of course.... We observed no small cells on our radar scope ... only a broad rain area."

A second report came from a departing four-engine airliner, which took off from Miami International at 1:18 p.m. and which utilized a departure pattern close to what Flight 705 would follow 17 minutes later. This crew described the worst turbulence as "medium to moderate" from just west of the airport to north of the Miami VORTAC. This aircraft was maintaining 5,000 feet at the time. The third report came from a crew in another 720B which was holding southeast of Miami at 13,000 feet. The CAB said this crew "observed numerous rain cells on radar in the Miami area and encountered light ice at this altitude."

The fourth crew interviewed was also flying a large jet. They "taxied out shortly after Flight 705 but delayed their takeoff for nearly an hour because of the weather."[1]

While these reports were being received, the Miami weather bureau radar observation (time-stamped as 1:44 p.m. showed "a pre-frontal squall line approximately 250 miles in length, oriented on a northeast-southwest line immediately northwest of Miami." The radar also indicated "a broken area of thunderstorms associated with this line, with cells two to twenty miles in diameter, and tops of detectable moisture at 30,000 feet. The line was moving southeast at eight knots, and moderate rain showers were occurring."[2] The CAB also noted that radiosonde ascents (weather balloon readings) made by the bureau at 6:00 a.m. and 6:00 p.m. that day showed that the freezing levels aloft were at 11,100 and 12,400 feet m.s.l. (mean sea level), respectively. The bureau's earlier forecasts were proving to be accurate.

Out in the Everglades, there were several witnesses to a loud midair explosion at around the time of Flight 705's last radio transmission. The CAB reported that six tourists on a fishing trip were among those. The group was fishing in Rookery Creek near a fishing hole called the "Little Banana Patch," in the upper reaches of the Shark River. Members of the group reported "that a loud explosion had occurred in the air, and several felt a subsequent ground tremor. They also reported that heavy rain had been falling in the area."[3] One woman in the group testified she heard "the sound of an explosion which had no echo. When she looked in that direction she saw an orange ball of flame in the edge of a cloud. As she directed the attention of her companions toward this flame, it dropped straight down, becoming a streak, and disappeared behind trees. Shortly after the disappearance a second sound was heard."

The group reported their location as at about seven miles south of where the explosion occurred. They immediately broke camp and headed towards the nearest marina in order to report the experience by phone. It took them three hours to reach the marina at Flamingo, Florida, and get in touch with the coast guard.

Meanwhile, ground controllers continued to attempt to raise Flight 705 on several different radio frequencies. Nothing was heard from 705 on the radio, nor did it show up at any airport in the area. Anxiety grew after 6:00 p.m. which is the point at which fuel onboard the 720B would have been exhausted. The hope that 705 was merely lost and having radio problems began to grow dim. Finally, at 6:45 p.m. some five hours after the crash and guided by eyewitness reports now coming in, a U.S. Coast Guard helicopter flown by Lt. Commander James Dillon of Akron, Ohio, spotted a bonfire in the distance. Hovering near, the chopper crew could tell they were dealing with a very large area of debris. Landing as carefully as they could in the gathering darkness, they looked around with flashlights and reported they found Flight 705. He radioed back to Miami: "There are no survivors."[4]

A ground team from the U.S. Fish and Game Commission reached the wreckage about midnight; they reported that some of the debris was still burning. Other searchers reaching the scene after daybreak that morning (February 13) discovered extremely fragmented remains of what was obviously a large jetliner spread over some 15 miles of the Everglades.[5] The crash site was not near any road or other landmark; officially it was marked as being at latitude 25° 33.9' N, and longitude 80° 53.0' W, some 40 miles or so southwest of Miami International Airport.[6]

The main part of the wreckage was located on fairly open and flat ground, "with outcroppings of coral rock, marshy water areas, and groves or hummocks of cypress trees irregularly spaced at one-half to one mile intervals."[7] The area was also filled with alligators, snakes and other wildlife. Recovery operations were difficult, to say the least. The nearest road of any kind was 15 miles away. Just getting to the site required over three hours of travel by four-wheel-drive vehicles and some 15 minutes by helicopter from the Tamiami Highway. Nonetheless, officials pressed on. Their first priority was recovery of the passengers and crew members. The site was immediately roped off and guarded by park rangers, Florida Highway Patrol officers, the Coast Guard and the FBI.

The CAB reported that many of the passengers remained in the largest intact part of the fuselage, some still strapped into their seats. The tops and sides of the passenger cabin had burned away.[8]

A temporary morgue was set up in a two-room schoolhouse at the Miccosukee Seminole Indian settlement, located 35 miles or so west of the airport on the old Tamiami Trail (US Highway 41). Students there were given an unexpected day out of school while their classroom filled with victims brought in by helicopter and truck. Officials worked from dawn until 10:00 a.m. to remove the bodies from the crash site. They were then transferred the following day (February 14) by hearse to the Dade County Morgue at Jackson Memorial Hospital.[9]

The first newsmen arrived at the scene the morning after the crash by swamp buggy, airboat and even chartered helicopter. They found government aviation experts beginning the painstaking work of what would be a long investigation. As the sun rose higher, the scope of what the investigators were facing became more apparent. The wreckage was distributed roughly in a line running from ENE to WSW. The area was approximately 1⅓ miles wide and 15 miles long, with around 90 percent of the debris, including large pieces, in the most westerly two miles. The other 10 percent was mostly light material blown to the east-northeast by prevailing winds aloft. Experienced investigators arriving on the scene could see immediately that an "in-flight breakup of the aircraft structure" had occurred prior to the jet hitting the ground.

At the furthest extent of the wreckage pattern was the upper part of the 720B's rudder. This spot was used by investigators as a zero point for plotting where individual pieces were found. Some 500 feet east of this zero point were engines one, two, four and three (in that order), which were found along a line from south to north a half-mile long. Around 500 feet northeast of the number three engine, investigators found the cockpit area. And approximately 1,500 feet east of the rudder fragment (the zero point) were the outboard portions of both wings. The main fuselage and wing center sections were found inverted 2,700 feet east of the zero point. This section of debris landed on a heading of 060 degrees, pointing to the northeast. The tail section was 1,000 feet farther east.[10] The investigation phase was underway.

The search for a probable cause of the loss of Flight 705 was the responsibility of Charles S. Collar, supervisory inspector of the Civil Aeronautics Board. On the scene in the Everglades as soon as he could make it after discovery of the wreckage, Collar immediately set in motion what was to prove to be a long and difficult probe. Besides effecting the care and removal of passengers and crew, Collar's team's immediate concern was finding the flight data recorder (FDR). Painted bright orange and able to withstand up to 100 Gs of force, the FDR would, they hoped, provide the answer to the crucial question of cause. They quickly located the battered instrument and sent it to experts at CAB headquarters in Washington to find out if useful data could be recovered from it.[11]

The team on the ground made an initial attempt to partially reconstruct the aircraft on the site, "but as the work progressed it became apparent that a more sophisticated study of the wreckage was required," so the team arranged for the use of a U.S. Coast Guard hangar at the Opa-Locka Airport in Miami for a complete reconstruction. A U.S. Army H-37 helicopter airlifted pieces of wreckage onto trucks or directly to the hangar.[12] Given the huge area where debris was scattered, and the extremely difficult conditions on the ground, an amazing 97 percent of the wreckage was recovered and taken to Opa-Locka. It took until April 1 to transfer, identify and reassemble the fragmented wreckage, but once this work was finished and detailed study resumed, a somewhat clearer picture emerged.

As the mockup took shape, investigators could see that the main failures in both wings and horizontal stabilizers were "in a downward direction, and virtually symmetrical. The forward fuselage broke upward and the vertical stabilizer failed to the left. All four engines generally separated upward and outboard."[13] Careful examination of the engines in particular revealed "certain peculiarities" regarding the number three engine in comparison to the others; it generated "considerable interest." Investigators reported that "the reverser on this engine landed approximately 1,300 feet from the main engine section." This engine bounced 150 feet away from its initial impact point in a north-northeast direction (015 degrees), whereas the other three engines bounced "approximately 40–45 feet on directions of 055, 080, and 060 degrees from their respective craters."[14]

The mockup also showed that "approximately four feet of the right wing ... and inboard of the [number three engine] nacelle, was broken away." There were indications that the engine's thrust reverser had collided with this section of wing.[15] It was an interesting anomaly. All four main engine mount fractures were examined for evidence of fatigue, "which might have resulted from damage sustained at the Fort Lauderdale accident" (of January 26, 1962), but no such evidence was found.

The mockup allowed the careful study of all flight control systems. Although it could not be absolutely determined that cables linking the cockpit with control surfaces were uncut at the time of breakup, "there was no evidence of any control system failure or malfunction except those associated with in-flight breakup or ground impact." In terms of electrical system fault and the possibility of hail or lightning damage, the findings were clear. There was "no evidence of arcing, burning, or electrical overload on any the generators. All available wiring bundles were examined for evidence of electrical arcing or beading. but none was found. There was no evidence of a lightning strike on any of the wreckage.... [The fuel venting system] in both wings was unobstructed and showed no fire damage. There was no evidence of internal wing tank fires prior to initial breakup. In addition, no evidence of hail damage was found on the nose section, or the leading edges of the wing, tail, or engine cowlings." The CAB even sent selected samples of the wreckage to the FBI's laboratory for detection of any explosive residue. The FBI found nothing suspicious.[16]

N724US was structurally sound at the time of its final flight—until something aloft tore it apart.

Prior to 1966, cockpit voice recorders (CVR) capable of capturing audio from several microphones located in an airliner's cockpit were not typically carried aboard U.S. passenger planes, although basic, early versions of them were available; N724US was not so equipped or required to carry a CVR. But it did carry a Fairchild flight data recorder, which was capable of recording in-flight pressure altitude, indicated airspeed, magnetic heading, and vertical acceleration as a function of time. Given the steady elimination of causes by examination of the mockup, the FDR would prove to be extremely valuable in the case of Northwest 705.

3. "Unusual conditions and circumstances" 33

Prior to the advent of CVR deployment in passenger aircraft, conversations between aircrews and ground controllers were recorded on the ground side, in control towers and en route centers. While valuable, they couldn't record what was happening in the cockpit. The FDR was therefore the only device on an aircraft which could provide crucial information in air transport accidents. Pictures of weather and flight radar were also recorded on the ground side. Once the FDR from ship 724 was opened and examined, officials were relieved that data could be recovered from its magnetic tape. That data told an invaluable story.

After takeoff on runway 27L at 1:35:22 p.m. Flight 705 made a series of turns to headings of south, southwest, west and northwest and climbed to 5,000 feet while experiencing light turbulence. At 1:42:46 p.m. clearance to a higher altitude was received and the jet rose; but it also encountered heavier turbulence during an approximately three-minute period. The FDR, primitive as it was, could detect turbulence by recording "large acceleration excursions," or wide variations in airspeed.[17] The larger the "excursion," the more severe was the turbulence.

At 1:45 p.m. 705 rose to 15,000 feet and turned left to a new heading of 200 degrees (south-southeast) before the heavy turbulence appeared to decrease. the indicated air speed (IAS), the speed of the jet in relation to the ground as shown on the cockpit instruments, during this period "fluctuated from 320 knots to 210 knots." At this point, the aircraft "continued climbing from 15,000 feet to 17,250 feet in a right turn which continued through 320 degrees, while the climb ceased and altitude remained constant for about 12 seconds." It was at this point that the FDR began to give up its secrets and paint a picture of an airplane in trouble, one in its last moments of life. At 1:47:25 p.m. the altitude began increasing again and the rate of climb gradually increased to approximately 9,000 feet per minute (three and a half times its normal rate) for about 13 seconds. Then the rate of climb suddenly decreased to 0 (the jet stopped rising), with the altitude peaking momentarily at 19,285 feet.

As the jetliner rose, the airspeed decreased, from 270 to 215 knots; the vertical accelerations changed rapidly from +1G to about –2G. Those aboard would have been pressed down into their seats at first, but would then suddenly find themselves floating up with a sensation of weightlessness. This negative G situation then fluctuated but ultimately reached "a mean value of about –2.8G." Even more alarming, the aircraft started descending increasingly fast. The FDR showed that G forces on the aircraft then changed from the high of –2.8G to +1.5G and then reversed again.[18] In other words, 705's occupants were slammed back down into their seats and then they floated back up. Anyone who didn't have a seat belt tightly fastened would undoubtedly have been

Civil Aeronautics Board composite photograph of the mockup of Boeing 720B N724US. The highly fragmented remains of Northwest Orient Airlines Flight 705 were reassembled in a hangar at Opa-Locka Airport to aid CAB investigators. The original photograph was heavily creased (Civil Aeronautics Board).

tossed around. (What happens in such a situation is shown in chapter 2 in the experience of Eastern Air Lines Flight 301 less than a year later.)

The CAB described how the end came: "In the last nine seconds of the readout the altitude trace continued to decrease, the air speed trace increased until the stylus hit the mechanical stop, the acceleration trace increased in a negative direction, and the heading remained fairly constant at 330 degrees." That the air speed exceeded the FDR's ability to trace it was a critical finding.[19] The jet had been tossed up into the air and had then fallen with great speed towards the earth. This final maneuver, from the onset of the climb at 1:47:25 p.m. until impact in the Everglades 19,000 feet below, lasted just

As funerals began for victims of the crash, and wreckage was helicoptered out of the Everglades, a large-scale investigation began, with a large number of institutions public and private working together and separately to pursue the crucial question: Was there something about the new jet aircraft, with highly swept wings, that made them more vulnerable in bad weather, especially at altitude? The accident possibly challenged the fondly held promise of "smooth flying above the weather." Answers were needed, on many levels.

From the first, the weather was the prime area of concern. The investigation called upon the facilities of the National Severe Storms Project (NSSP), a project of the air force, the FAA and NASA (and a handful of other alphabet-soup agencies), which had been formed to "study the formation and life history of squall lines." NSSP director C.F. Van Thullenar testified at the public hearing. His testimony revolved around the nature of thunderstorms and turbulence,[20] which would be critical not only in the case of Northwest 705, but also other accidents in the 1960s. He told the board that "the turbulence

3. "Unusual conditions and circumstances" 35

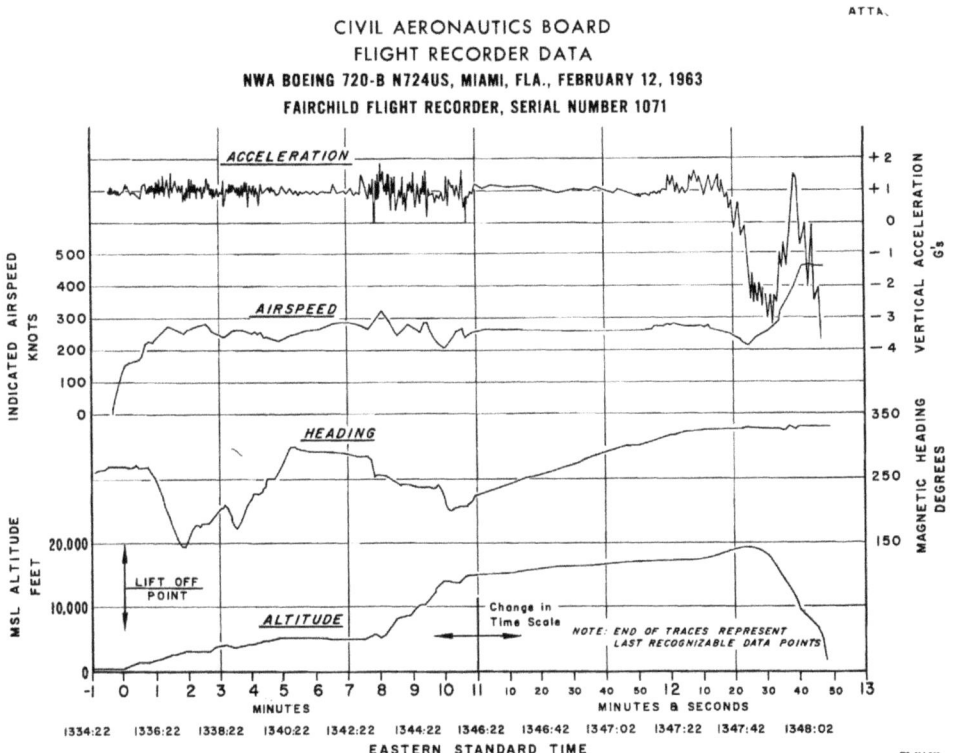

Northwest Flight 705 flight data recorder traces from the Civil Aeronautics Board. Data collected from the point of departure from Miami International Airport is particularly dramatic at the end of the tape. Notice how the aircraft falls from almost 20,000 feet into the Everglades in less than 30 seconds (Civil Aeronautics Board).

encountered in a thunderstorm varies directly with the amount of rainfall and the diameter of the storm during its building or mature stage. During the deteriorating stage, the diameter of the storm is no longer indicative of the turbulence." He also testified that updrafts, such as the one probably encountered by Flight 705, can be quite large, "frequently 15 miles wide" and "invariably contain smaller gusts which produce the turbulence." He added that "the strength of these smaller gusts generally varies directly with that of the draft in which they occur." The larger the updraft, the more severe the turbulence within it.

In addition to Van Thullenar's testimony, the CAB also received a report from the NSSP that went into further detail; the report "concluded in part that it is not unreasonable to assume that severe turbulence exists at some point in any storm, and in a growing, or large mature thunderstorm one may expect extreme turbulence." The report was in the hands of the investigators by June of 1963, some four months after Flight 705 went down.[21]

The U.S. Weather Bureau itself was also called upon for analysis and testimony. The bureau's job was to evaluate the "nine indicators of turbulence which might have been present in the crash area at the time of the accident." The USWB's conclusions echoed those of the NSSP. The most reliable of the nine indicators "seems to be the rainfall rate, which indicates gust values in the severe range, other fairly reliable indications such as buoyancy, hail, and surface gusts indicated somewhat higher gust values."[22] Given the weather data recorded on February 12, 1963, it was reasonable to conclude that the turbulence Flight 705 encountered aloft was severe to extreme.

During accident investigations, the CAB (and now the National Transportation Safety Board, or NTSB) has a team of investigators assigned to evaluate "human factors"; and testimony about the people involved in an accident, from the cockpit to the control tower to the en route center and, sometimes, to the airline board room, is usually a feature of formal hearings. In Flight 705's case, investigators were particularly interested in what the three cockpit crew members had experienced, and they also zeroed in on how the turbulence encountered would have possibly affected Captain Almquist, who was the pilot flying at the time of the incident (as mentioned above, First Officer Feller was handling communications duties).

At the formal hearing, testimony was given from an unnamed representative of the Naval Medical Research Institute (NMRI), as well as a professional pilot who performed as the representative's subject "during a series of tests on negative G maneuvers conducted by the U.S. Navy at their Johnsville, Pennsylvania facility." (These tests were performed using the center's centrifuge; it would not be the last time it was used in this period to understand the effect of turbulence on humans.) Both of these witnesses testified that "from a physiological standpoint," the accelerations recorded on the FDR "should not have physically incapacitated the crew members," as long as they were restrained in their seats, which was indeed the case.

The navy tests involved subjecting the pilot test subject to "repeated –3 G loads for periods of up to 30 seconds, and –5 G for shorter intervals." The pilot testified he experienced "no adverse physiological effects." The NMRI representative added that the forces used in the testing had been duplicated in real-world flights and in centrifuge testing. Their testimony included a caveat, however: "If one has never been exposed to high negative G forces, the experience could be frightening." Whether they thought that Captain Almquist or his crewmates, veterans of thousands of hours of flight in a wide variety of conditions, could have been "frightened" was not recorded by the CAB.[23]

The second area of focus for the investigation was the aircraft itself.

Boeing, the 720B's manufacturer, conducted two studies which were focused on three issues:

    1. The capability of the aircraft to perform the maneuver indicated by the flight recorder readout;
    2. What control inputs would be required; and
    3. What aircraft response would result from partial or complete loss of the horizontal tail.

Boeing first used an analog computer study, which provided data for a more sophisticated IBM digital simulation. As part of the simulation, both of these studies tested "varied longitudinal control inputs," i.e., different pitch positions, from nose up to nose down. This was done to try to duplicate the data seen on the FDR.

The simulations showed that the FDR's data could be re-created but that putting the aircraft into a full nose-down using the horizontal stabilizer and elevator at their furthest limits was required to produce the dive made by N724US. The elevator would also have to be intact to make the partial recovery shown on the FDR. If the horizontal tail surfaces had failed fully or partially, the dive would have been steeper and faster. They were therefore probably intact as the dive began. The Boeing studies also showed that at the beginning of the final 45-second maneuver, the nose was pitched up 22 degrees; during the final dive, it was angled more than 90 degrees down, or past the vertical. It would have been highly disorienting, to say the least.

The manufacturer also conducted a study to compare the climb performance recorded by Flight 705's FDR and what was known about the 720B's climb capability in general. Any deviation in N724US's capability from that of normal generic 720Bs would show one of two things: (1) If there was a loss in airspeed and a gain in altitude, an updraft would have been encountered; or (2) If the opposite was the case, a downdraft would be present. This comparison showed that "drafts of high intensity were acting on the aircraft at the time of the high rate of climb [i.e., when the nose pitched up 22 degrees], and during the dive." These up and down drafts were not shown to be severe enough to cause damage to the aircraft structure, however.

Another crucial Boeing study simulated flights of a 720B to show how it reacted to various up or down drafts. Flights were simulated where the pilot did not seek to control the aircraft and just let it "ride out" the drafts. Other flights had the pilot use "sufficient control" to maintain level flight. The flights where the pilot let the plane ride out the drafts resulted in less acceleration forces. The aircraft stabilized itself and was not upset by the forces it encountered. This would be a crucial lesson for the future.[24]

For a number of years, pilots had known to reduce speed when turbulence was encountered. They were usually provided with charts which delineated the airspeed to maintain for the specific plane they were flying when it encountered turbulence. Airlines also specified procedures in addition to airspeeds. As the CAB noted, Northwest's turbulence penetration procedure was "typical of the industry." There was also training for "recovery from unusual attitudes, not exceeding 10 degrees in pitch with 45 degrees of bank," which was a part of the training and testing of 720B captains. But upset recovery procedures were not required of Northwest pilots until they received their type rating in the aircraft. It was "never repeated in recurrent training nor was any flight simulator training provided."[25]

Northwest Airlines itself conducted various tests, both on its own and in conjunction with Boeing and with the Batelle Institute, a nonprofit research and development organization. The airline particularly was concerned with a phenomenon known to affect the 720B called "balance bay icing." The elevator balance bay area of the aircraft is an accessible area where the cables connecting the elevators to the control column can be adjusted and maintained. There had been 13 instances of pitch difficulty due to icing in this compartment that restricted elevator movement and effectiveness, although none of them caused a loss of control. In those 13 cases, the difficulty was resolved by either a descent to a lower altitude or by "greater than normal pilot inputs."[26]

Northwest and Batelle performed flight tests in conditions similar to those of February 12; and they concluded that, while ambient temperatures in the balance bay were around 40°F, the phenomenon of "cold soaking" might have frozen the elevators and caused Captain Almquist to attempt to unfreeze them by moving the control column as in a few of the recorded incidents above, with "greater than normal inputs." The Northwest-Batelle study, released in December of 1964, almost two years after the accident, asserted that N724US's balance bay, via the cold-soaking phenomenon, would have reached the freezing mark temperature-wise just before the final, fatal maneuver. Cold soaking is described by NASA in current instructional literature regarding aircraft icing:

> Ice can form even when the outside air temperature (OAT) is well above 0°C (32°F). An aircraft equipped with wing fuel tanks may have fuel that is at a sufficiently low temperature such that it lowers the wing skin temperature to below the freezing point of water. If an aircraft has been at a high altitude, where cold temperature prevails, for a period of time, the aircrafts' major structural components such as the wing, tail and fuselage will assume the lower temperature, which will often be below the freezing point.... While on the ground, the cold soaked aircraft will cause ice to form when liquid water, either as condensation from the atmosphere or as rain, comes in contact with critical surfaces.[27]

The Civil Aeronautics Board considered the balance bay/cold soaking theory put forward by Northwest-Batelle but rejected it. Investigators noted that the study itself demonstrated that ambient temperatures were above freezing, and they offered a rebuttal of cold soaking in the particular case of Flight 705:

> In developing their thesis that temperatures in the balance bay area were substantially below the freezing level, the report presents no new weather evidence, but rather it presents a different interpretation of the evidence considered by the Board in its analysis. The Board did not find persuasive their "cold-soak" reasoning, their assumption of a 20-degree differential between rain and ambient temperatures, and their method of determining the temperature variation with altitude in the accident area. In the absence of a more conclusive showing that the structural temperature in the balance bay area were appreciably below the freezing level, the main Northwest-Batelle conclusion that immobilization of the elevators early in the climb precipitated the large longitudinal control displacements is without substance.[28]

The board did note, however, that much of the material in the Northwest-Batelle report, specifically "the flightpath analysis, the significance of the long down-elevator period, the human factors influences," concurred with its own findings.

During the two-plus years of the inquiry made into the crash of Northwest Flight 705, a massive amount of work was undertaken. A full discussion of all these tests and studies and simulations involved is beyond the scope of this particular narrative; it is undoubtedly enough for an entire book on its own. While some disagreement among the professionals occurred along the way (including about the final conclusions of the board), that work did result in a reasonably clear picture of what happened in the air the afternoon of February 12, 1963. The board did not believe it was "necessary or possible" to be extremely precise about the final 45 seconds of Flight 705, but investigators did believe a generalized picture emerged, "sufficient for determining a probable cause and for providing a clear understanding of the general problem."

Those final 45 seconds, said the board, started shortly after 1:47 p.m. when the aircraft entered an area of severe turbulence. The aircraft climbed, either as a result of updrafts or pilot input, but probably a combination of both. This resulted in "rapidly decreasing airspeed, increasing rate of climb and a high nose attitude" that would cue any pilot to "take drastic action to prevent what would appear to be an impending stall."[29] The pilot flying, Captain Almquist, probably acted on this situation while being subjected to "severe vibrating accelerations from the turbulence," vibrations that probably made his instruments difficult to see. He then used full down elevator as well as aircraft nosedown (AND) stabilizer trim in tandem to get out of the situation. This combination of two actions had the effect of arresting the drop in airspeed and the high nose-up pitch, but it would have produced "extremely high negative G forces on the aircraft." Those G forces, along with the aircraft's elevator control characteristics, according to the board, explain why a successful recovery was not made.[30]

The situation in the cockpit would have been chaotic in "any airliner with a crew totally unaccustomed to forces of this type and magnitude." Warning lights and ringing bells were going off, any loose items in the cockpit were tossed around, and the flight crew were tossed up against their seat restraints to such an extent that an average airline pilot "would probably have difficulty keeping his feet on the rudder pedals and his hands on the control wheel."[31] This conclusion led the board to believe that it was "inconceivable to believe that the pilot continued to apply full down elevator" during the high G force period. Instead, the board asserted, it was much more reasonable to believe that "elevator control forces lightened" (in a manner consistent with National Aeronautics and Space Administration (NASA) wind tunnel tests conducted during the investigation), and when that happened, with the pilots' hands probably taken off the control wheel, the control column would have remained in "full forward or nosedown position."[32]

It then appeared that Almquist "managed to place his hands on the control wheel some eight seconds later," but the aircraft at that point was in a vertical dive at 16,000 feet with airspeed rapidly increasing. The FDR indicated that the elevator returned to neutral for a few seconds and then moved to a full up position. "By this time the airspeed was at or beyond 470 knots, the altitude was nearing 10,000 feet, and the vertical acceleration" was slowing, meaning excessive airspeed and air loads made a successful recovery impossible. While the aircraft dove through 10,000 feet, the pilot attempted to retrim the stabilizer to the noseup position, but aerodynamic forces stalled the motor driving the elevators and they did not respond.[33]

Boeing's testing posited that a successful recovery could have been made at around 14,000 feet if airspeed was below 320 knots. But in view of the actual conditions onboard Flight 705 (not to mention the extreme human force required), it would be "unreasonable to fault the crew for not being able to accomplish such a recovery. In fact, it was entirely possible that recovery attempts made by rapid upward elevator movements by the crew might have made the elevator and horizontal tail surfaces fail even earlier."

The blurring of a pilot's view of instruments which were not designed for high readability in such conditions was a contributing problem, according to the board. "The cockpit acceleration environment induced by fuselage bending response in heavy turbulence," along with the bouncing up and down on the seat in such weather, was "an unfavorable characteristic of all large, swept-wing transports."[34] If a pilot took his eyes off the attitude indicator (which shows whether a plane is level in pitch and roll) for even a fraction of a second, it could be difficult to impossible under the circumstances for him to refocus on it, especially under the highly stressful situation confronting the crew. The board implicated the HZ-4 attitude indicator installed in N724US as part of the problem. It was "adequate, although by no means optimized" for normal or near-normal pitch attitudes. But during high angles of attack, interpretation of the instrument was "extremely difficult because the horizon reference line on the indicator recedes from the face of the instrument," away from a pilot's face. In the final dive and possible recovery attempt, this could have made a difficult situation even worse.

As for Captain Almquist specifically, the board believed him to be qualified and "possessed of average or better flying abilities" according to the standards of the time. However, the board believed that a "clearer understanding of the 'limits' of an 'average' pilot must be found." It suggested that statistical methods could be applied to prescribe a "realistic capability range" for the so-called average pilot which would help manufacturers design an optimally safe aircraft.[35]

Finally, the board concluded that Boeing's recovery study indicated a recovery was (perhaps theoretically) possible, and investigators "preferred to avoid stating that a successful recovery could not have been made although there are some reasons to believe this latter possibility is more nearly correct." The board then drove the point home: "In any event there is no intended implication that the pilot did not do everything possible to regain and maintain control under the most unusual conditions and circumstances."[36]

"Unusual conditions and circumstances" is an understatement. The reality was that mere seconds were available to effect a recovery. And within those few seconds, the overstressed aircraft began to break apart. The board's studies showed that as the plane dove through 10,000 feet the tail plane probably failed first, falling to the left; this was quickly followed by the stabilizers and wings bending and failing downwards. The engines and a large section of the forward fuselage (the cockpit and a section of the forward cabin) were then also split off, flying upwards at first. The remaining hulk of the aircraft kept falling straight down. Since the aircraft was at the beginning of a thousand-mile trip, the aircraft's fuel tanks were almost full. When the 720B broke up at 10,000 feet, this fuel vaporized into the air and ignited, creating the fireball seen by witnesses on the ground.

Over two years of hard work and study by a near army of expert investigators, scientists, engineers, pilots and others would come down to this: Flight 705 was upset during severe to extreme turbulence; it was *theoretically* possible to recover from the upset, but such a recovery was not made in the seconds available. The resulting forces broke the aircraft into pieces, ignited the fuel aboard, and killed 43 people.

The board's final report takes note of the effort of the entire aviation community's "considerable attention and effort to the upset problem, and that many, real safety changes in today's operations have been brought about as a result"[37] of this effort. The FAA's efforts to educate pilots to potential hazards of turbulence, along with many safety bulletins, airline training programs and the U.S. Weather Bureau's expediting of weather radar displays on air traffic control scopes were just a few of these efforts, all of which are laudable and important. But even if change was as fast as the jetliner itself, it would prove not to be fast enough. While the investigation proceeded and changes were introduced, more and more jetliners carried more and more passengers and crews aloft into conditions not then understood. And at least two more jet upsets, along with another crash disturbingly familiar, were to follow fast on the heels of Northwest Flight 705.

Northwest Airlines Flight 705

Summary
Northwest Airlines, Inc.
Boeing 720B, N724US
In the Everglades, 37 miles west-southwest of Miami International Airport, Florida
Tuesday, February 12, 1963
Fatalities: 35 passengers, 8 crew; no survivors

Report
Civil Aeronautics Board
Aircraft Accident Report
SA. 372, File No. 1–0006
Adopted June 1, 1965; released June 4, 1965

Fatalities

Crew
Capt. Roy W. Almquist, 47, Rosemont, MN
First Officer Robert J. Feller, 38, Minneapolis
Second Officer Allen R. Friesen, 29, Hopkins, MN
Stewardess Wendy Engebretson, 22, Minneapolis
Stewardess Connie Rae Blank, 21, Minneapolis
Stewardess Myrna A. Ewert, 25, Minneapolis

Stewardess Virginia Lee Younkin, 25, Minneapolis
Stewardess Mary S. Sandell, 20, Minneapolis

Passengers
Baldwin, Henry, Tacoma
Broman, Signe, Chicago
Cain, Joseph E., Indianapolis
Christianson, Mrs. Wilbur L., Minneapolis
Christianson, Wilbur L. Minneapolis

Diaz, Ramon, Cuba
Enloe, George A., Seattle
Galler, Dan, Chicago
Goodwin, Mrs. Ted, Grand Rapids
Goodwin, Ted, Grand Rapids
Heil, John C., Seattle
Hollerich, Jack P., La Salle, IL
Irwin, W.R., Omaha
Kelinson, M.D., Portland, OR
Kelinson, Mrs. M.D., Portland, OR
Lebodow, Mrs. Fanny, Lincolnwood, IL
Melahn, Arnold, Cary, IL
Melahn, Mrs. A., Cary, IL
Nimsch, Fred, Caracas, Venezuela
Olson, Fred III, Rockford, IL
Olson, Joan, Rockford, IL

Orszule, Jerilyn, 20, Cicero, IL
Orszule, Mrs. Walter, Berwyn, IL
Orszule, Walter, Berwyn, IL
Orszule, Walter Jr., 18, Berwyn, IL
Rand, A.B., Kennilworth, IL
Rever, Christine, Rockford, IL
Rhea, Gilmore, Deerfield, IL
Rhea, Mrs. Gilmore, Deerfield, IL
Schwendener, Susan, Rockford, IL
Smigiel, Mrs. Anton, Niles, IL
Srodulski, Mrs. Rose (Joseph), Park Ridge, IL
Tengerstron, E.W., Chicago
Tengerstron, Mrs. E.W., Chicago
Wells, Dr. H.E., Chicago
Wubbold, Joseph, Coral Gables, FL

4

"A turbulence problem"? Persistent Questions

We don't know if we have a turbulence problem or not.
—CAB director Bobbie R. Allen

It may have qualified as one of the most surreal moments in the life of a British tourist, a man who later admitted to having started his vacation a tad early by "having a few" while still an hour or two away from landing in Mexico City. The man suddenly found himself glued to the roof of a jetliner 20,000 feet over Texas while his wife tried to pull him down into his seat by the back of his trousers.

The wife later told reporters that all her husband could think to say was, "Let go my trousers woman—you're pulling them off."

He managed to keep his trousers on and his dignity for the most part intact, even when he and 20 other passengers slammed back down into their seats as the flight crew of Eastern Air Lines Flight 301 managed to save the DC-8. But the state of his pants would be the least of his worries for the next two hours.[1]

While investigators from the Civil Aeronautics Board spent most of 1963 examining, testing and discussing the inflight breakup of Northwest Orient 705, more and more passengers took to airways around the world. And more and more of them were being carried in the new jet transports. By the end of 1963 the International Civil Aviation Organization (ICAO) would record a record 134 million passengers carried by all airlines in the group's 101 member states. And since the Soviet Union and People's Republic of China were not members of ICAO, the actual number of passengers worldwide was probably much higher.[2] The Northwest 705 investigation was therefore crucial to ever-increasing millions of passengers, especially since officials admitted that they weren't exactly sure if they had a "turbulence problem" or not, much less possible solutions.

Najeeb Halaby was the director of the Federal Aviation Agency between 1961 and 1965 and the future father-in-law of King Hussein of Jordan. In a display of candor which showed his no-nonsense training as a military test pilot and skydiver, he told *Life* magazine, "I can discern neither a pattern nor a panacea, and so I don't sleep well nights. This is a terrible admission to make, but I wouldn't know what to do about it if we did prove turbulence was the cause. I guess we think that the system of corrections we've launched will be the answer."[3] That kind of direct admission is so unthinkable in today's world that it's almost hard to believe Halaby actually said it. Yet, his misgivings and frustration were echoed by other professionals of the time, in equally candid admissions to *Life* associate editor Warren R. Young.

Newton Lieurance, director of Aviation Weather Affairs for the U.S. Weather Bureau, told Young what the bureau knew about turbulence at the time: violently agitated air existed within thunderstorms and jet streams, and there was "shearing action" between those streams and other air that produces "swirling eddies." He said (somewhat ominously, given what would happen to Braniff Flight 250 three years later) that there were also vertical waves of rolling air.[4] (Those vertical waves came off the lee side of mountains, he said; it would take Dr. Ted Fujita's Braniff 250 investigation to set that story straight—that vertical rolling waves can be found in places other than the mountains.) Lieurance added, "Now we've had to change our minds. We don't even know what types exist. We really don't have a satisfactory theory for turbulence and it looks as though we won't have for years to come."[5]

The parade of worried officials in Young's excellent *Life* piece ended with the Civil Aeronautics Board director, Bobbie R. Allen, who indicated in the summer of 1963 that all the turbulence talk might just be "hot air." Young makes no comment on whether Allen was joking or not. "If I were trying to write a best-selling novel, I'd put 'turbulence' in the title—there seems to be so much interest," Allen told *Life*. (So noted, Mr. Director!) But his last sentence would prove to be more on the mark—and more sobering: "We don't know if we have a turbulence problem or not."[6]

By November 1963, the board's report on Northwest 705 was still not ready. (It would finally be released in the summer of 1965.) Yet more jetliners were leaving the factories at Douglas, Boeing, Convair and Lockheed (the "L-188 Jet-Prop Electra," in the latter's case) to carry more people greater distances at faster speeds, all at higher altitudes. The ICAO reported that the average number of passengers per flight being carried had increased from 17 people in 1948 to 42 in 1963.[7]

Investigations of crashes prior to the Jet Age had developed fairly sophisticated systems and processes to figure out whether man, machine or environment or some combination of them caused the event. The development of this process had resulted in a situation where most dangers had "by now been eliminated from modern air transport,"[8] wrote *Life*'s Warren R. Young. But he also noted that airlines in the U.S. in 1962 carried 71 million people 50 billion miles, but only 121 were killed during those operations. The year 1963 would see an 11 percent increase in the number of passengers carried on jets.[9]

Young repeats the old adage of the industry: the car ride on the road to the airport is the most dangerous part of a trip. Air travel in the Piston Era had developed to the point where taking a bath in your own tub was more dangerous than taking a DC-7 to see Grandma in Dubuque. But the Jet Age presented so many new prospects, good and bad, in its enticing "more people higher, faster, and further" equation. So many unknowns were up there at those speeds and heights that it wasn't possible to know about them until they were encountered. Yet "holding your breath and whistling through a graveyard" (to mix two old clichés in an impossible action) is never a workable response to an unknown and potentially serious threat.

The experts, the airlines, the government, with plenty of breath-holding both on the ground and aloft, had to answer an urgent two-part question: (1) Is there a problem with the new Jet Age aircraft? and (2) If there is, what do we do about it? While trying to stay calm, methodical and careful in order to get the answer right, those involved in the Flight 705 investigation were very aware of the important role the investigation was playing in those two key questions. In the year after Flight 705 went down, there would be several incidents nationwide which would press the experts for speedy answers and make the public ever more aware of the situation. And one of those incidents would be even more deadly than the loss of Flight 705.

5

"It broke the sound barrier at least once": United Airlines Flight 746

Without warning, the plane hit the turbulent air and dropped, jostling the passengers.
—UPI Press Pool Report, July 13, 1963

---

*Boeing 720B
Over O'Neill, Nebraska
10:30 p.m. (approx.), July 12, 1963
Friday*

On July 12, 1963 (five months to the day after Northwest 705 impacted in the Everglades), a United Airlines Boeing 720B jetliner (similar in most aspects to Flight 705's 720B) departed San Francisco International Airport (SFO) as scheduled at 6:25 p.m. PDT.[1] Destination: Chicago's O'Hare International Airport (ORD), with an estimated arrival time of 12:10 a.m. CDT. Flight 746 was commanded by Captain Lynden E. Duescher, a 22-year United veteran with wide-ranging experience. The 720B, with 56 passengers and a crew of seven on board, headed east over the Sierra Nevada and the cabin crew began serving breakfast.

Capt. Duescher, 42, hailed from Duck Creek, Wisconsin, and attended the University of Wisconsin. He joined United in 1941, working all the way up the flight ladder. During his first four years alone, he progressed from student pilot to first officer to first pilot to cargo captain and finally to captain. In 1951, United gave him a diamond-studded lapel pin for his first 10 years of service, and in 1961 another for his 20-year anniversary. He was an experienced and capable airline captain.[2] He was joined on the flight deck by two other experienced flyers: First Officer Eric Anderson, 32, of Elmhurst, and Second Officer Ervin A. Rochlitz, 41, of Oak Lawn. They were also fortunate to have another experienced pilot, E.P. Aiken, who was flying from San Francisco to take up his duty rounds in Chicago. Anderson possessed 10,200 hour of flight time in his log book; Rochlitz had 10,000.[3]

United's summer 1963 timetable shows Flight 746 "operated by B-720 'Boeing 720 Jet Mainliner' with 'JET One-Class Service' and cocktail and meal service appropriate to the time of day." Like Northwest's marketing, United also highlighted the comfort and high class of service and meals aboard. They had been calling their aircraft "Mainliners" since the early days of the DC-3 in the 1930s. The timetable explains how Jet Mainliner flights worked for passengers. Seat selection was made at check-in. The one-class fare from SFO to ORD was typically $111; on two-class planes, fares were $135.40 for first and $105.45 for coach. (According to *Measuring Worth*, an online site that calculates relative values of prices over a number of years, $111 in 1963 would cost around $815 in 2012 dollars.[4] There is also a side note of historical curiosity: Between 1954 and 1970, United operated so-called Executive Flights—for men only—between Chicago and New York, for just $3 over the regular fare. The all-male passenger complement was served by an all-female stewardess crew. The flights featured cocktail service, steaks, business magazines and cigars.[5] Even though it was quite popular, changing social conditions meant the day of Executive Flights was effectively over.)

Onboard United 746 flying was normal. Meal service was over and most passengers were asleep or relaxing with after-dinner drinks. But as the 720B approached Nebraska, the flight crew could see a squall line out ahead. After some discussion, the crew decided to get above the probable turbulence in the towering clouds and received clearance for a climb to FL410 (41,000 feet). The "fasten seat belts" sign was illuminated and the cabin crew gave warning before the aircraft hit rough air.[6]

Anderson, the pilot flying at the time, overflew O'Neill, Nebraska, at 37,500 feet and continued climbing to the new altitude. Just above O'Neill, Flight 746 was suddenly caught in a violent down gust. The big Boeing nosed over and hurtled towards the earth with gut-wrenching effect. As the surprised and alarmed crew dealt with spinning instruments and the turbulence inside the cockpit which made it difficult to see (just as the final report described the situation in the Northwest Flight 705 cockpit), the aircraft reached 450 knots and beyond; neither the cockpit altimeter nor the flight data recorder could record those high speeds. It was calculated later that the 720B broke the sound barrier at least once.[7]

Anderson wrestled with the controls and Duescher added extra muscle. Fortunately, the Boeing managed to stay intact, the crew using wing spoilers and flaps as well as reduced power to arrest the steep descent. A shallow dive let them recover before they hit the ground; they had fallen almost 23,000 feet in just a minute and a half.

One man was not belted in, but he hung onto seat rests and managed to remain uninjured in the process. Given that the aircraft appeared intact and operating normally and that the facilities and help needed were at Chicago O'Hare, Flight 746 pressed on. The air smoothed out and landing was normal. Once the plane was on the ground, medical treatment was given to six passengers and crew members who were shaken up. The aircraft was towed away. Investigators—well aware that United 746 was almost an exact copy of Northwest 705, albeit with a happier ending—pulled the flight data recorder for analysis and started the usual investigation. The aircraft was inspected minutely and eventually returned to service.[8]

A little over a year after the United 746 incident, the FAA made a request of the Aviation Medical Acceleration Laboratory in Johnsville (AMAL). The AMAL, part of the U.S. Naval Air Development Center (under the direction of B.F. Burgess, Jr., CDR, MSC, USN), had the largest human centrifuge in use at the time, part of the development of space and the race to land a man on the moon. The FAA asked the lab to "utilize the human centrifuge to investigate factors contributing to commercial jet aircraft crashes associated with severe air turbulence." The crashes included those of Northwest 705 and Eastern 304 and the near-crashes of United 746 and Eastern 301.

The center's research director, Carl F. Schmidt, M.D., put a team to work on the request. Stuart Ragland, Jr., CDR, MC, USN; Randall M. Chambers, PhD; Richard J. Crosbie, MA; and Lloyd Hitchcock, Jr., PhD, worked on the final report, "Simulation and Effects of Severe Turbulence on Jet Airline Pilots,"[9] and delivered it to the FAA August 13, 1964. The team used the experience of United Flight 746 and tried to answer two specific questions: (1) "Whether or not an adequate simulation of the physical events taken from the flight recorder of a United Airlines Boeing 720-B could be reproduced with the centrifuge," and (2) "[I]f there were any effects upon pilots under these circumstances detrimental to the safe control of the aircraft."[10]

AMAL's team successfully programmed and simulated United 746's experience with the centrifuge. This simulated turbulence "produced accelerations that fluctuated from a maximum of $+3.5Gz$ to a maximum of $-2Gz$ at a random frequency average of 1 cps (cycles per second)."[11] Interestingly, United 746's flight crew, Captain Duescher and First Officer Anderson, were put into the centrifuge for the first test of the simulation. AMAL reported: "They pronounced it excellent." (Whether they had "flashbacks" or other problems during their Flight 746 replay was not recorded. Presumably, they had no problems.) In addition to Duesher and Anderson "eight other airline pilots [two of them FAA test pilots] were exposed to the simulation and adjudged it realistic based on their personal experiences in turbulence during their pilot careers."

During the simulations, which AMAL described as limited, staff "observed and recorded" some pilot performance degradation which was "detrimental to safe control" of an aircraft. Specifically, researchers reported there appeared to be a consistent tendency to experience a kinesthetic illusion which causes the pilot to make inappropriate pitch control movements." In a description which would contribute to the understanding of the crash of Eastern Flight 304, researchers recorded how the pilots reacted to certain situations in the centrifuge: "When negative Gz was encountered for the first time, an initial movement of the yoke in the wrong direction was the rule rather than the exception. Some stick movements that were thought to be involuntary resulting from jostle were made, but these were not considered to be of a magnitude sufficient to hazard normal aircraft control."[12]

One thing that helped the pilots was "the use of a shoulder harness as well as a secure lap belt," which "made control easier and made the pilots feel more secure psychologically." The pilots themselves reported "some blurring of the instruments." This could be countered. "If the pilot concentrated upon the artificial horizon, he could maintain his orientation with regard to that instrument but was unable to maintain a useful panel scan."[13] In other words, the pilot could hold the aircraft steady only by boring in on one instrument while ignoring all the others. In turbulence, with the other instruments spinning and flashing and klaxons and warning bells ringing, such concentration was, to put it mildly, unlikely.

The testing included different kinds of attitude indicator instruments: "All pilots felt the Lear 3-inch model 4003 G, Type MM3 artificial horizon used in the simulation was easier to interpret than the type instrument employed in their commercial jet aircraft, which may become unreadable in unusual attitudes."[14] The likelihood of airlines replacing their existing indicators with the Lear model was also, to put it mildly, unlikely. Physiologically speaking, the testing revealed that the turbulence experience exhausted the pilots quicker than "normal instrument flying." Even though each simulation lasted only six minutes, the airmen were drained by the strain of trying to control the situation—a situation which had occurred in real life aboard United 746. There was some good news: "Disorientation was not a prominent feature in this experiment and motion sickness did not occur. No abnormal physiologic responses were encountered."

The researchers summarized their findings for the FAA: "These data strongly suggest that by responding to a strong kinesthetic illusion of climb or dive after correcting from an unusual nose up or nose down attitude pilots are creating ever increasing deviations from normal flight pitch attitude in both directions alternately, somewhat analogous to pilot-induced oscillations, until the aircraft stalls and falls off into a steep dive that is difficult to recognize or to recover from because of the limitations inherent in the types of artificial horizons frequently employed in their aircraft."[15] What this finding was describing had happened at least twice already in the period while the report was being prepared (fall 1963 to summer 1964). This use of simulation to help investigators grasp how an incident or accident unfolded was pioneering. As hardware became smaller and more powerful and software more capable, simulation would become an increasingly important tool available to accident investigators over the years.

When United Flight 746 landed in Chicago that July night, the nation's air transport system had dodged yet another bullet. But those in the industry were very aware there had been a razor-thin difference between Northwest cratering into the Everglades and United landing safely in Chicago. Answers such as the AMAL simulation were needed faster than ever, since more and more jets of different types were taking to the air.

Summary

United Air Lines, Inc.
Flight 746
Boeing 720-B

Over O'Neill, NE
July 12, 1963
Fatalities: 0. Injuries: 6

6

"The Nation's Most Progressive Airline": Eastern Air Lines

> *Eastern offers more jet service between the North and major cities in the South and Southwest than any other airline! Eastern also jets to Montreal, Bermuda, Puerto Rico and Mexico.*
> —Eastern Air Lines timetable, fall 1963

Over at Eastern Air Lines, one of America's pioneering airlines, the Jet Age began January 24, 1960, with a Douglas DC-8-21 flying the company's premier route, New York to Miami. Eastern used its 15 "Diesel Eights" initially on its longer routes, including those to Mexico and Central and South America.[1] Eastern's larger-than-life chairman, Eddie Rickenbacker, called his new liners "Golden Falcon Jets," continuing a tradition where he named the airline's Martin 4-0-4s the "Great Silver Fleet," and the Douglas DC-7s and Lockheed L-1049 Constellations the "Golden Falcons."[2]

The Golden Falcon Jet DC-8s were, like the Boeing 707/720 series, jets with 30 degrees of sweep back. They required a different flying approach than the slow propeller equipment in the fleet. For the first three years, as Eastern's crews built up time and experience in the cockpit, the airline's DC-8s performed beautifully. Southbound Flight 301 from New York (Idlewild) nonstop to Mexico City and the return flight, 304—which was a night owl-type of service back to New York with stops in New Orleans, Atlanta and Washington Dulles—were typical of the airline's use of the new jet. Passengers loved the speed and quieter cabins.[3] (*Flight* magazine in 1960 noted the high noise levels in the DC-7C, a four-engined, propeller aircraft in wide use in the 1950s, saying the "relatively high interior noise and vibration levels ... resulted from the 3,250 h.p. Wright compound engines in an airframe designed originally (as the DC-4) for units of 1,450 h.p."[4] With the need to reduce this impact and jet power plants becoming available, the time was ripe for the next step in the Douglas company's series: an all new, jet-powered, DC-8.

The DC-8 was announced in 1955, first flew on May 30, 1958, and achieved its certification on August 31, 1959. It entered service with both United Airlines and Delta Air Lines on September 18, 1959. Douglas had 145 initial orders from 20 different airlines, but the actual production run would last 14 years, feature seven major variants and turn out 556 aircraft. According to the Boeing Company, more than 300 DC-8's remained in service as late as 1995, with more than 340 flights scheduled per day. Sixty-nine of those remained in service as of 2010.[5]

The first Diesel 8s off the line had the following dimensions/capabilities:

- 142'5" wingspan
- 150'6" length
- 42'4" high
- 355,000 pounds maximum weight
- 570 mph
- 35,000-foot ceiling
- 4,773-mile range
- Four 13,500-lb. PW-JT3C turbojets
- Three flight crew members, eight flight attendants and 117–259 passengers seated in several available layouts.[6]

The DC-8 was the first commercial transport to break the sound barrier, which occurred during a test-flight dive. It was taken for granted during this period that transports like the DC-8 would be somewhat stop-gap aircraft, ending the piston/propeller era and leading directly and quickly to supersonic transports, which would then be the dominant aircraft for commercial transport. So breaking the sound barrier and proving the durability of the design and construction was an important milestone to reach. A Douglas company dispatch dated Wednesday, August 21, 1961, announced the important supersonic milestone, which occurred over Edwards Air Force Base, California:

> Earlier today, during a routine certification test flight, Douglas Chief Pilot Bill Magruder flew the aircraft faster than the speed of sound, making the DC-8 the first Commercial Jet Transport to break the sound barrier. After climbing to an altitude of 52,090 feet, the DC-8–42 series aircraft attained a maximum speed of Mach 1.012 or 660 mph while in a controlled dive through 41,088 feet. The purpose of the flight was to collect data on a new leading-edge design for the wing. ... Upon completion of testing, the record setting aircraft will be delivered

to Canadian Pacific Air Lines for regular scheduled service. In June, the DC-8 set three speed records while being operated by Delta Airlines.[7]

The obvious durability and high capability of the aircraft were thus fully proven as far as Douglas was concerned.

The Douglas Aircraft Company produced a film chronicling these early days of their new baby. In *Birth of a Jet*,[8] Douglas noted that the rigorous DC-8 testing program included 130,000 flights for gust load, landing, takeoff, and underwater testing; five years of wind tunnel tests; and structural proof tests.

The jet's April 19, 1958, rollout after six years of R&D and testing, at Long Beach, California, took place in front of dignitaries, from congressmen to airline executives to stewardesses representing airline customers to Douglas employees and founder Don Douglas himself.

After the rollout ceremony, ground and flight test programs ensued for over a month. On May 30, the jet departed Long Beach, needing less than a third of the runway to get airborne. Two hours of cruising over southern California followed, and then the jet landed smoothly at Edwards Air Force Base for further testing. Nine DC-8s were in the test program; some of them were used to train the airline crews who would soon operate them.

Douglas and the airlines were proud of the technical achievements of the DC-8 on the exterior, but they especially talked up what was going on inside. The sophisticated first-class cabin would have between 118 and 122 seats; tourist class could hold up to 144 seats. The jet was the size, said Douglas, of a three-bedroom house with a den. Air scoops below the nose provided cooling system without fumes; the system was so powerful it could simultaneously cool nine normal homes. The engines were muffled with, among other things, "daisy" nozzles on the rear of each engine, which added to noise abatement and a quiet interior environment. Douglas promised that the comfort onboard would be enhanced by the fact that flights of less than half the time of piston airliners would become accessible to everyone.

And while passengers were aloft, they would have large "sky view windows" to watch the landscape far below speeding beneath their feet. The aircraft could fly as high as 40,000 feet, but the pressurization systems would feel like the altitude at 6,700 feet. (Fifty years on, this goal is still at the forefront of airliner designers' thoughts: Boeing's 787 Dreamliner, just going into service at this writing, has promised to have better views and better atmosphere inside with higher humidity, which can minimize jet lag and other effects of high-altitude travel. Only time will tell if this 55-year-old promise has finally been delivered.)

As the DC-8 was tested and underwent its first airline service, test pilot A.G. Heimerdinger of Douglas sang the aircraft's praises in an issue of *Flying* in June 1959. Heimerdinger was a Douglas engineering test pilot and flight operations manager. He especially noticed the 8's stability: "There have been many pleasant surprises in the DC-8. Airplanes with sharply sweptback wings often have a 'Dutch roll' characteristic. The DC-8 has 30 degrees of sweepback, yet we have never even hooked up the yaw damper which the engineers perfected. When we cut an outboard engine at cruising with our feet off the rudder, the nose swings only three or four degrees to one side."[9] Heimerdinger also noted the extensiveness of testing: "We put a complete cabin section under water and ran it through 140,000 cycles at maximum pressure differential; that's equivalent to 100 years of steady flying. In another instance we fired harpoons at the fuselage, attempting to make it rip in explosive decompression. Use of titanium rip stops, however, limited the size of the puncture to a small hole which allowed pressure to leak out slowly and did not damage the structure."[10] In the view of its test pilot, the DC-8 was ready to ride the airwaves.

The DC-8 entered service with Eastern Air Lines a few months after United and Delta put their Series 11s and 12s in the air. Rickenbacker decided to wait the extra time for Douglas to produce the Series 21. There would be 15 DC-8–21s in Eastern colors in the first few years (it would eventually have 43 of all types in its fleet; the final Eastern 8s were Series 61s and they served until late December of 1982).

The Series 21 for which Rickenbacker was willing to wait (and thereby give Delta the distinction of flying the DC-8 for the first time in commercial service) featured JT4A engines providing between 15,800 and 17,500 pounds of thrust, compared to the Series 11 and 12, which had JT3C engines with 13,500 pounds. The 21s also had higher maximum weight capabilities and had extended wingtip and leading edge slots. The better engines no longer needed the water injection required by the JT3Cs.[11]

Even so, it was becoming apparent that Rickenbacker was aging and seemed behind the curve; he had been resistant until it was almost too late to even consider buying jets in the first place. As more and more jets joined the Golden Fleet, Rickenbacker was removed as CEO on October 1, 1959, and fully retired as chairman at the end of 1963. America's World War I Flying Ace was gone, but other aces of the air were about to replace him as heroes in their own right.

While aviation pioneer and ace Eddie Rickenbacker is synonymous with Eastern Air Lines, it was actually a different air pioneer/ace who founded Eastern and set it on a path it followed practically to the end, which was an ignominious bankruptcy and shutdown in 1991. Harold Pitcairn, like Rickenbacker, was born in the 1890s and would achieve success on the Western Front in World War I. While Rickenbacker was born in poverty, Pitcairn was the son of one of the founders of Pittsburgh Plate Glass, now known as PPG Industries, and was accustomed to wealth. At 18, he completed flight training and operated seaplanes out of the Glenn Curtiss school in Buffalo, New York. He worked for his father's company and then as president of the Owosso Sugar Company in Michigan (his father and others at PPG had invested in Owosso and Harold served as their representative on the board).[12] When the United States joined the European war in 1917, Pitcairn left the family businesses and joined the Army Air Service for training. The armistice on November 11, 1918, meant that he was too late for the fighting in France, but his training certificate was signed by one of the two ultimate pioneers of flight, Orville Wright.[13]

With the war over, Pitcairn preferred to remain in aviation rather than return to glassmaking. In 1926, the United States Postal Service opened its air mail routes around the country to competitive bidding by private companies. Pitcairn was successful in bidding for the New York to Atlanta route. The only obstacle was that the contract he signed provided for a flight schedule that no existing aircraft could fulfill. The Postal Service wanted Pitcairn's operation to put up to 600 pounds of mail on an aircraft that would leave Atlanta at the close of the business day and arrive in New York in time for the first postal delivery of the next day.

While at army cadet school, Pitcairn became friends with Agnew Larsen, an aspiring engineer. All through the 1920s, Pitcairn and Larsen informally experimented with a variety of aircraft and "autogiro" (early helicopter) designs. After Pitcairn obtained the New York to Atlanta mail contract, he persuaded Larsen to formalize the relationship; Larsen became chief engineer of the Pitcairn Aircraft Company. The company's design efforts resulted in a series of Pitcairn biplanes (called the Mailwing series) capable of the speed and capacity the air mail contracts demanded. Each number of the series improved on the performance of the previous aircraft. Pitcairn's efforts were successful enough that he also bid on the Atlanta to Miami route; his bid was accepted and Pitcairn Aviation, the line formed to operate the planes built by Pitcairn Aircraft Company, began running mail and passengers up and down the East Coast.

Pitcairn's interest was always more focused on the invention of the "autogiro," the early rotary-wing/vertical-lift aircraft created by Spanish inventor Juan de la Cierva. Pitcairn and his company initiated dozens of patents for all sorts of aviation-related improvements, held by the Pitcairn-Cierva Autogyro Company of America, which began licensing the technology in 1929. So convinced was Pitcairn that vertical lift was the future of commercial aviation that he decided to sell his airline with its air mail contracts so he could concentrate on the Autogyro.

These efforts were rewarded in 1931; Pitcairn was awarded the Collier Trophy, the highest award in American aviation, for his vertical lift craft. Pitcairn then went one better; he landed an autogyro on the South Lawn of the

White House and accepted the trophy in person from the hand of President Herbert Hoover. By that point, Pitcairn and the East Coast airline that bore his name were each going in different directions. In 1929, the owner of North American Aviation, Clement Keys, purchased Pitcairn Aviation and its mail contracts and renamed it Eastern Air Transport, to reflect that Pitcairn was no longer associated with the line and the fact that the company had reach all up and down the eastern seaboard.

Pitcairn's sale to Keys was mostly good timing. In addition to the normal chaos in the pioneering and infant commercial aviation industry, there were plenty of other obstacles to be overcome in the period. The stock market crash of 1929, the change in administrations and the coming of Franklin Roosevelt's New Deal after 1933 would bring external pressures; but Keys pressed on. The airline survived and was sold profitably to General Motors in 1934.[14]

Shortly after the sale, reform-minded representatives in Congress decided the Postal Service was guilty of collusion with the airlines in the way routes were originally divided up. Other pioneers starting their own carriers needed to be able to compete for the contracts. Congress punished the airlines involved by banning them from carrying mail and ordered the army to resume flying the mail as it had until 1926. The results were disastrous. In the years since the original contracts were awarded, the airlines became experienced at dealing with all the en route hazards involved in flying in all weather and across all terrain at all times. The army no longer had that expertise, and crashes and pilot deaths and large-scale economic loss followed. Congress ended the short experiment by ordering the Postal Service to let new contracts to private carriers, once again pulling the Army Air Service out of the business. But lawmakers stipulated that the original colluding airlines, such as Eastern Air Transport, could not take part in the bidding.

For GM and other airline owners, the problem was easily solved: just change the airline's name. Eastern Air Transport merged with a company named New York Airlines and GM rechristened the whole "new" operation Eastern Air Lines, the name it would keep until its dissolution in 1991. Two other companies would soon join it: Ludington Air Lines and Wedell-Williams Transport.

In 1938, Eddie Rickenbacker bought Eastern from GM for $3.5 million and began a modern new era. Eastern contributed, like every other company in the U.S., to the war effort between 1941 and 1945, then experienced, like other airlines, tremendous growth and profit in the postwar period. Rickenbacker's exploits from World War I to the beginning of the 1960s have been exhaustively reported elsewhere, so it's enough to record here that, where Eastern was concerned, he professionalized and shepherded an ever-growing company, one of the largest airlines in the world, into the Jet Age.

It was Rickenbacker, said to be hesitant about the new jets, who nonetheless made the decision to order Douglas DC-8s. As mentioned earlier, while those aircraft were being built Rickenbacker was ousted as day-to-day CEO in October 1959, and completely separated from the company in 1963, just as the DC-8s were taking over more and more of Eastern's routes. The "Diesel 8s" were joined in 1962 by Boeing 720s and in 1964 by Boeing 727s. Eastern's Jet Age was well underway by the winter of 1963–64.[15]

The growth that Eastern experienced in the post–World War II period saw its Golden Falcons and Great Silver Fleet flying from Canada to Mexico and South America by the 1960s via interchange agreements with airlines such as Braniff. Their routes took them as far west as Chicago, Minneapolis, St. Louis, Dallas and Houston and as far east as Boston, Montreal, Bermuda and San Juan, Puerto Rico. As at Northwest, routes to Florida and Mexico were popular with Eastern's customers. On July 23, 1957, Eastern began a New York-New Orleans-Mexico City service using DC-7 aircraft. DC-8s were put on the route as the new jets came off the assembly line. Eastern Flight 301 would leave New York's Idlewild Airport (renamed Kennedy after 1963) at 10:00 a.m. and arrive in Mexico City, after flying nonstop, at 12:37 p.m. (times are local to each airport). Eastern also offered routings with connections through either Atlanta or New Orleans, with arrival in Mexico City at 9:45 p.m.

Northbound Flight 304 was a different situation. Flight 304 left Mexico City at 10:45 p.m. and operated as an overnight service with stops in New Orleans, Atlanta, Washington-Dulles, and New York Idlewild. It reached New Orleans at 12:47 a.m., Atlanta at 3:40 a.m., and Washington-Dulles at 7:05 a.m., and was back in New York at 8:07 a.m.[16] An interesting side note is that fares in late 1963 and early 1964 for Eastern's passengers in first class travel between New York and Mexico City were listed at $167.30 (in 1963 dollars). That amount is equal to $1,230 in 2011 dollars. Tourist class fares were $118 in 1963 dollars, the equivalent of $867 in 2011 dollars. Since Eastern no longer exists, I compared the 1963 fares with those charged today on the same route by AeroMexico, which has a similar schedule to Eastern's 1963 timetable. As of early 2013, AeroMexico was charging $1,379.10 for a one-way, first class fare from New York Kennedy to Mexico City. That's the equivalent of $188 in 2011 dollars—$70 more than Eastern was charging for Tourist class in 1963.[17] And in those days before "load management" techniques and perfected algorithms and massive computer power, each passenger paid the same fare as those sitting in the seats around her. In some ways, the 1978 deregulation of airline routes and fares did not make cheaper fares a reality.

It was to be these two Eastern flights, 301 and 304, which would give investigators, regulators, and the industry more nightmares about jet upset in turbulence in the winter of 1963–64.

7

A Strange, Weird Thing: Eastern Air Lines Flight 301

I could see visually this heavy dark area to my left. On the right I could see visually a small dark area. In other words, it looked like a [thunderstorm] cell over on the right. However, my airplane radar did not show any echo for the cell on the screen.
—Eastern Air Lines Flight 301 Captain Mel French,
testimony before the CAB accident investigation board

Douglas DC-8–21 N8603
Newport, Texas
10:00 a.m. (approx.), November 9, 1963
Saturday

On a November Saturday morning, 121 passengers waited in Eastern Air Lines' Terminal 1 at New York's Idlewild Airport to board that day's Flight 301 to Mexico City. The flight crew of three and cabin crew of four were ready to go, but the aircraft was not. Mechanics were called in to repair an unspecified fault in the flight control system, which caused a two-hour delay.[1] Some passengers must have chafed at the delay to the start of their Mexican vacations. It was a dreary New York late autumn day: rainy/drizzly with overcast skies, 57 degrees and a 13-mph wind. Mexico City reported clear skies and a high temperature of 68 degrees.[2]

Eastern's Flight 301 operated that day with a DC-8–21 registered N8603. Built in Long Beach on November 24, 1959, (the 37th DC-8 off the line) and delivered to Eastern February 14, 1960, ship 603 served the line's longer routes.[3] Equipped with four JT4A-9 engines, the Series 21 jets could serve "hot and high" airports (the higher the altitude the thinner the air, which affects aircraft performance) such as Mexico City. They had room for 132 passengers in first and coach classes. Flight 301 was under the control that day of flight veterans Captain Mel H. French of Huntingdon Station, New York, and First Officer Grant R. Newby of Manhattan. Both were long-time pilots and had ample experience in the DC-8.[4]

By noon, the fault was repaired, the aircraft was loaded and it was sent on its way. The delays however, were not over. As the flight proceeded southward near Texas, it ran into storms and headwinds, which both slowed it down and caused deviations from the planned flight track.[5] These added up to a higher-than-planned fuel burn rate, which in turn meant a stop for fuel would be needed. Captain French made the call to land at Houston International Airport (now named William P. Hobby Airport) to refuel. This was accomplished as quickly as possible and Flight 301 was soon airborne again. There was some "grumbling" among the passengers, but they settled down as the "no smoking" and "fasten seat belts" signs were turned off and things returned to normal. Captain French announced the aircraft was climbing smoothly to its assigned 31,000 feet and that Mexico City was now just an hour and 40 minutes away.[6] Those in the cabin began to relax as the stewardesses on board began making their rounds.

But nothing about Flight 301 that day was destined to be relaxing for long. As the flight approached 20,000 feet just 45 miles southwest of Houston, Captain French, apparently warned by ground controllers receiving other pilot reports, announced that seat belts should be refastened. In the next few seconds, Flight 301 entered rough air; it hit an air pocket and then simply dropped like an elevator out from under the 128 people on board.[7] Those who had either ignored Captain French's warning or had no time to react found themselves in "odd" positions. One man found himself on the ceiling of the aircraft, his wife, who was belted in, hanging onto the bottom of his trousers. There were other passengers up there with him, too taken by surprise to scream. Since aircraft at the time did not have closed bins with secure latches, everything that was in the baggage racks on top began flying around with the passengers. Handbags fell open, scattering lipsticks and face powder, mirrors and money, all over the cabin. Ashtrays which had been in use since departure from New York sprang open and ashes and cigarette butts sprayed the air with a fine gray mist. For several seconds, it was like being in zero-gravity in outer space.[8]

In the cockpit, this leg of the flight was being flown by First Officer Newby, while Captain French handled communications. Captain French told Civil Aeronautics Board investigators later that they were climbing "into an arch of clouds between two thunderstorm cells and heading toward the blue" portion of the sky, when the aircraft encountered the leading edge of the turbulence.[9] They "hit a bump like driving across a railroad track," said French, and then had a hard time believing their eyes. The air speed indicators on both the captain's and the first officer's panels swung rapidly down to zero. Newby immediately reacted by pushing the control column forward in order to regain the lost airspeed by lowering the nose. But the DC-8 just "swooped over like a roller coaster" and entered an uncontrollable, high-speed dive. The altimeters unwound quickly as they lost more and more height. French later estimated they had dropped about 13,000 feet in 20 seconds or less.[10]

Finally, Captain French decided on a somewhat risky maneuver; he activated the thrust reversers on all four engines and slowed the dive enough that a recovery became possible. Such maneuvers can result in structural damage, even in extremely well-built and tough aircraft. And N8603 was no exception. With both pilots straining on the control column, they managed to level off the jetliner at just 5,000 feet. But as it leveled off, one of the passengers, a Philadelphia doctor named M.C. Elroy, testified, he looked out and saw the wing violently vibrating, followed by the inboard number three engine suddenly ripping away from its mounting and falling backwards in a flash. As it fell, it hit and dented the tail of the aircraft and then dropped into a pasture between Damon and Newgulf, Texas, southwest of Houston.[11] In the passenger cabin, those 20 or so who found themselves glued to the ceiling now dropped back towards the seats as quickly as they had risen, along with all the luggage, miscellaneous articles and cigarette ash, which made some passengers look as if they had been in a volcanic eruption.

By this time, French and Newby had the airliner back under control and level at 6,000 feet. It seemed to be stable and flyable, even without its number three engine.[12] They did need to get back on the ground, but there were obstacles. First, the runways back at Houston were probably not long enough to handle the damaged aircraft with its full load of passengers, cargo and fuel. (Houston Intercontinental, with its wider and longer runways and better terminal, was in the planning stages and would not debut until 1968.) Second, all that fuel that had been added in Houston to get them to Mexico City was now a hindrance, an irony not lost on the crew.

Captain French reviewed all the options with ground controllers and checked on the status of the passengers. Told there were injuries, but none life-threatening, French made his command decision: they would head for Barksdale Air Force Base near Shreveport, Louisiana, 260 miles back to the north. This would allow time to burn off or dump the now-unnecessary fuel and would give them access to the longest runways available in the area, as well as extra help from the air force. Barksdale's longest runway was 11,756 feet long; Houston's was just 8,000. French pointed the wounded DC-8 towards Shreveport.[13] To the passengers, he said, "We don't know the exact situation, but we seem to be under control up here."[14]

Flight 301 remained in the air for another hour or so. The passengers were prepared by the cabin crew for a possible crash landing. Many of them had no idea that one of the four engines was gone. Some prayed. A truck driver, certain he was about to die, still managed to make a joke about the missing engine; but he made his seatmate nervous and decided to keep quiet. All the passengers were impressed by Eastern's crew, noting how calm and professional they remained throughout the experience.[15]

Barksdale scrambled all of its available resources. When Flight 301 finally arrived overhead and circled a few times to burn off more fuel, six ambulances, six fire trucks, 10 doctors and 25 hospital corpsmen were standing by, along with many others ready to help out. Barksdale also sent up a fighter jet; its pilot conducted a visual survey of the DC-8, particularly the landing gear as it lowered. As the sun began to set in Shreveport, Flight 301's some six hours of drama finally came to an end. The flight crew eased the DC-8 onto Barksdale's runway in a manner described as "silken smooth." In fact, the landing was the only thing that had gone right all day.[16]

A press pool photo taken as passengers exited the aircraft at Barksdale shows what looks to be a normal arrival scene: people walk down a stairway and head for the terminal, and none show any outward sign of their ordeal.[17] But 17 of them were injured, the most serious a broken back. An air force officer who observed the arrival told investigators, "They walked out white as ghosts. Some of them were gray with cigarette ashes. A few were mussed up, but all were calm."[18] Those who were basically uninjured and in no need of hospital treatment were taken to the base officers club, where they were treated to dinner and discussed their experience. Eastern quickly dispatched a replacement DC-8 to Barksdale. The replacement aircraft departed the air base with 100 of Flight 301's original passengers. They finally arrived safely in Mexico City, somewhat worse for wear.[19]

In Shreveport, CAB investigators soon arrived to inspect the aircraft and open an inquiry. Captain French told them the turbulence was the worst he had encountered in his long career. The flight and cabin crews were given medical examinations and leaves of absence.[20] Newby was restored to flight status on November 21, the day before the assassination of U.S. president John F. Kennedy.[21]

While CAB investigators added Flight 301's near disaster to their caseload (and to their sense of urgency over the jet upset issue), the Federal Aviation Agency limited its action to "advice to air crews" on handling jets in turbulence. The agency recommended that existing guidance on jet turbulence penetration speed be altered. Established procedures in 1963 called for jetliners to be slowed from their normal cruise speed in the neighborhood of 550 mph down to 300 mph when rough air was either suspected or experienced.

Between the time Northwest Flight 705 crashed in February and the near-catastrophic experience of Eastern 301 in November, the agency changed its mind and revised the recommended penetration speed upwards to 325 mph.[22]

The agency also had a warning where autopilot was concerned: pilots should make sure the autopilot's "altitude hold" was off. Altitude hold tells an autopilot to maintain a certain altitude. In the case of a sudden drop like that experienced by United 746 or Eastern 301, the autopilot could attempt to hold the aircraft at, say, 20,000 feet of altitude by raising the nose higher and higher to regain that flight level. This could in turn put the aircraft into a stall or upset situation or both.[23]

Both of these warnings had their roots in the experiences of Northwest 705 and United 746, which caused deaths and injuries. But notice that the agency used the terms "recommendation" and "guidance." The learning was not put into a firm regulation with the force of law. Airlines were free to follow the guidelines or ignore them, if they were willing to assume the risk. In speaking directly to pilots, the FAA seemed to make an appeal over the heads of the airlines. But pilots had no more incentive than airlines to follow the suggestions.

Meanwhile, turbulence continued to upset jets. The CAB reported that an American Airlines Boeing 707, flying above Alamosa, Colorado, hit severe turbulence about which its captain, H.D. Schmidt, testified: "I can't honestly say that a person was in control of the airplane.... You don't get panicky but a man would be an idiot to say you don't get scared."[24]

Postscript, of sorts: Eastern's daily operation of Flight 301 between New York and Mexico City resumed the next morning. But high drama would find the flight one more time. On November 13, 1963, just four days after the turbulence drama over Texas, Eastern's run to the border was interrupted again. This time it wasn't turbulence in the air, just a telephone call claiming a bomb had been placed on that day's Flight 301. The pilots turned the jetliner, another DC-8, back to Idlewild, where an extensive search was performed. No bomb was found, and the aircraft finally touched down in Mexico City several hours late.[25]

Summary

Eastern Air Lines, Inc.	Newport, TX
Flight 301	November 9, 1963
Douglas DC-8-21, N8603, S/N 45428	Fatalities: 0. Injuries: 20

8

Disappeared into the Overcast: Eastern Air Lines Flight 304

Shattered bodies, suitcases, chunks of wreckage, a child's first grade reader bobbed in whitecapped waves. Tides swept debris towards the Gulf of Mexico.
—UPI Press Pool Report, February 26, 1964

Douglas DC-8-21 N8607
Lake Pontchartrain, near New Orleans, Louisiana
2:02 a.m., February 25, 1964
Tuesday

On the night of February 23, 1964, southbound Eastern Air Lines Flight 305, captained by veteran William B. Zeng, with Grant Newby once again in the first officer's seat, arrived in Mexico City from Atlanta and New Orleans. Flight engineer Harry Idol was the third crew member. They had been on duty for eight and a half hours, which had been uneventful. The crew was due to helm Flight 304 back north the next night.[1]

First Officer Newby, after the wild ride on Flight 301 back in November, was back into the routine of flying the line. As mentioned above, he was given a two-week leave after the jet upset over Texas. His experience since then had been comfortingly normal. He passed a DC-8 proficiency check on December 4, 1963, and then a first-class medical examination on January 28, 1964. Since the November upset, he had performed first officer duties on 20 separate trips totaling 214 hours of flight time. By the time he arrived in Mexico City, Newby had flown 2,404 hours in the DC-8; his log books showed that he had flown 10,734 hours as a commercial pilot.[2]

8. Disappeared into the Overcast

Eastern Air Lines DC-8-21, N8607, which crashed into Lake Pontchartrain near New Orleans, Louisiana, while operating Flight 304 between Mexico City and New York (Eddie Coates Collection).

Captain Zeng, 47, of Ringoes, New Jersey, was also an experienced veteran. While he had less time than Newby on the DC-8 (916 hours), his total pilot time (19,160 hours) was almost twice that of the first officer. Zeng had type ratings stretching back years to the DC-3, -4, -6/7 and -8; the Martin 2-0-2/4-0-4 series; and the Lockheed Constellation series and L-188 Electra "prop-jet." Where the DC-8 was concerned, he received his initial rating on January 8, 1962, and his latest proficiency check on January 24, 1964. Zeng passed his first-class medical exam on August 27, 1963.[3]

Harry Idol, 39, of South Farmingdale, New York, was not only a flight engineer, but also held a full airline transport pilot certificate, rated for the Martin 202/204 series. He had 8,300 hours on his pilot's log, 1,069 of it on the DC-8 as pilot or engineer. Idol's first-class medical certificate was dated September 23, 1963.[4]

The trio mostly rested during their 24-hour layover in Mexico City. They were scheduled to operate northbound Flight 304, a "red-eye" type of flight scheduled for nightly departures at 10:45 p.m. Unlike the nonstop, southbound Flight 301, the northbound 304 had three scheduled stops in New Orleans, Atlanta, and Washington, D.C. (Dulles). The flight would reach New York's newly renamed John F. Kennedy International Airport at 8:07 the next morning.[5]

On the evening of February 24, 1964, Eastern's N8607 was the DC-8 operating the southbound Flight 305. Just like the previous night's flight that Zeng, Newby, and Idol operated, 305 on February 24th was mostly uneventful. The captain (whose name was not recorded in CAB documents) reported on arrival that "the only exception to normality was that the pitch trim compensator (PTC) was inoperative, with a fix scheduled for the next morning at Kennedy Airport." This information was noted by Captain Zeng as he prepared for the return to New York. As a result, Captain Zeng filed an instrument flight rules (IFR) flight plan for Flight 304 to be operated at a reduced airspeed, which was standard Eastern Air Lines procedure for flying a DC-8 with an inoperative PTC.[6]

When preflight activities were complete and passengers loaded, Flight 304 pushed away from the gate almost on time. The first segment was operated with four flight attendants who were scheduled to deplane in New Orleans and be replaced by a new group for the Atlanta portion. The attendants who left the flight in New Orleans were later interviewed by investigators and noted two things. First, Captain Zeng was at the controls and planned to switch flying duties with First Officer Newby on each segment; and second, the last thirty minutes of the trip were marked by light to moderate turbulence during the approach and landing phases. Flight 304 landed at Moisant International slightly behind schedule, at 12:51 a.m., February 25, 1964.[7]

Normal turn-around activities followed. The flight crew checked weather and inspected the aircraft, the manifest of passengers and flight, and calculations of weights and balances. They were advised of the U.S. Weather Bureau's aviation area forecast, which was to be valid from 1:00 a.m. to 1:00 p.m. This forecast indicated "a surface wave off the Louisiana coast, expected to move eastward at 30–35 knots, with ceilings at 400–800 feet and moderate to occasionally heavy rain. A north-south line of showers and embedded thunderstorms north of this surface wave was expected to produce moderate to severe turbulence in the thunderstorms and heavier showers, and moderate or greater clear air turbulence was forecast from 24,000 to 40,000 feet," throughout the greater New Orleans area.[8] In addition, the crew had Eastern Air Lines' own system forecast, valid from midnight to noon. The Eastern weather dispatchers predicted "ceilings below 1,000 feet, light rain in the Pensacola–New Orleans area." Turbulence was also forecast at moderate to severe levels in thunderstorms, along with "light to moderate wind shear turbulence above 14,000 feet."[9]

Approximately 13 minutes before Flight 304 departed, the weather bureau's radar showed an area of scattered echoes containing light rain showers, with the closest showers located 60 miles west-southwest of the radar location. The bureau also issued a surface weather observation some ten minutes after

304's takeoff, which showed 1,000-foot ceilings in overcast conditions, visibility on the surface at seven miles and winds from the north-northeast at 12 knots. The closest observation of winds aloft was taken at Burrwood, Louisiana, 70 miles southeast of New Orleans. It showed wind velocities of 34 to 57 knots between 3,000 and 9,000 feet, and a freezing level aloft at 12,400 feet.[10]

While the flight crew prepared for the next segment, the U.S. Customs Service inspected the aircraft and baggage, since it had arrived from a foreign destination. The four fresh flight attendants came aboard with 51 passengers, 34 of whom were continuing on from Mexico City, and 17 of whom were originating in New Orleans. Fourteen of the 51 passengers were Eastern Air Lines employees riding on company passes. With everything buttoned up and passengers and crew in their places, Newby taxied the aircraft to runway one for departure.[11]

There was a brief pause while 304 waited for a Curtiss C-46 twin-propeller aircraft to depart. This slower aircraft was using a similar flight path as the DC-8 would use, so plenty of time was allowed for separation. The C-46's crew reported moderate to severe turbulence which began immediately on takeoff and lasted until they reached 9,000 feet. They also reported that airspeed fluctuated 15 to 20 knots and the heading varied approximately 10 degrees. The C-46 maintained control, however, and the rest of its flight was uneventful.[12] As soon as the C-46 was out of the way, First Officer Newby advanced the throttles and the DC-8 gathered speed. Another large jet turned onto runway one as 304 rotated off the pavement. The captain of this jet testified later that Eastern 304 disappeared into the overcast at just 1,200 feet. It was 1:59 a.m.[13]

An air traffic controller in the New Orleans Moisant tower directed Flight 304 to contact departure control at 2:02 a.m. Captain Zeng acknowledged the instruction, since First Officer Newby was flying the aircraft. The controller also saw Flight 304 disappear into the overcast at an estimated two or three miles north of the airport. A departure controller then took over, telling 304 to "turn right, heading 030, be a vector north of J-37,"[14] the planned routing to Atlanta.

The departure controller then issued instructions for 304's next contact, New Orleans Air Route Traffic Control Center (ARTCC). This center identified the DC-8 on radar as being five miles north of the airport's VORTAC. Finally, 304 was instructed to "contact New Orleans Center radar, frequency 123.6 now." At 2:03:15 a.m., the crew replied, "OK." It was to be the final transmission from Eastern Flight 304.[15]

Two minutes later, the New Orleans Center and the departure controllers conferred and noted that no further transmissions had been received and that the DC-8's radar signature had disappeared from their scopes. Emergency procedures were then initiated. Flight 304's last known position was some eight miles from the airport, over Lake Pontchartrain. At the same time, 29 witnesses in various locations on or near the lake saw a fire-like glow in the vicinity. At least 11 of the witnesses "reported hearing an explosive rumble," and three of them described "a tornado-like sound or terrible scream."[16]

9

Extraordinary Lives: Grant Newby, Marie-Hélène Lefaucheux and Kenneth Lee Spencer

> *The status of women is still at different degrees and still, practically, inferior to that of men throughout the world.... There is reason to believe that if women were to be allowed an equal share of public responsibilities with men the world would be more stable and more civilized.*
> —Marie-Hélène Lefaucheux, remarks at the official opening of the International Council of Women, Washington, D.C., June 20, 1963

The alarm over Eastern 304 raised by the controllers was answered by a variety of first responders, including the U.S. Coast Guard, which sent out a helicopter to the last reported position. This crew spotted an oil slick and floating debris in the lake. "This search rapidly assumed enormous proportions as additional electronic and sonic underwater detection gear became available," CAB investigators noted.[1]

At the point where the debris was spotted, Lake Pontchartrain was reportedly 16 feet deep, but searchers could see that much of the remains of the aircraft was deeply embedded in the lake's mud, which was estimated later to be 20 feet thick.[2] On the surface of the lake, the coast guard reported it found "seats, communication equipment and soundproofing ... clothing, luggage, and what was described as parts of bodies." They also found a child's first grade reader.[3] Officials also said "parts of the debris, including insurance papers, were definitely linked with at least two passengers on the Eastern plane."[4] At a period when several aircraft had been bombed by passengers attempting to effect large insurance payouts (which would also be the plot of a 1968 book by Arthur Hailey, *Airport*),⁵ reporters were looking for the most likely explanations for the crash.

The recovery operation intensified as the sun rose. At dawn, a coast guard boat retrieved the first body of a victim, wrapped up in one of the plane's yellow life rafts. The service told reporters that other bodies were found "torn and broken." Divers began feeling their way in extremely murky and hazardous conditions along the bottom of the lake.⁶ As is usual in most crashes, there were initial hopes that survivors might be located. But dawn quickly dashed those hopes. Debris bobbing on the lake was too fragmented. This accident was "non-survivable."⁷ Meanwhile, officials had a different immediate concern; some of the debris was floating towards the Gulf of Mexico on the tide.⁸ It needed to be recovered quickly.

The Civil Aeronautics Board's William L. Lamb arrived from Washington, D.C., to head the investigation. There were initially 20 investigators on the scene beginning what would be a very long process. Eastern's own crisis teams were on hand, and reporters converged from all over the country. The airline's New York office released a list of passengers and crew members during the afternoon of February 25.⁹ Reporters who were responsible for covering the industry almost immediately noticed a name familiar from just three months previously: that of first officer Grant Newby. They also took note of two other somewhat famous names on the passenger list: Kenneth Lee Spencer and Madame Marie-Hélène Lefaucheux.¹⁰

The former was a celebrated bass singer with a long and storied career which included highly praised roles in two Hollywood movies of the World War II years, *Cabin in the Sky* and *Bataan*. Spencer was initially an understudy for Paul Robeson in *Showboat*, and then performed the role himself to wide acclaim. During the previous few years, Spencer was performing operatic roles in Germany and other countries. He was on his way from performances in Mexico to another tour of Germany.¹¹ Marie-Hélène Lefaucheux was a prominent member of the French Resistance during the war, after which she served in several roles in French politics and became a member of the nation's delegation to the United Nations. She was one of 15 founding members of the U.N.'s Commission on the Status of Women. Her experiences in the French underground were partially portrayed by Leslie Caron in the 1966 movie, *Is Paris Burning?*¹²

Before getting into the somewhat complicated details of the investigation, it's important to first put a human face on the story, since technical discussions can blunt or even obliterate the tragedy, pain and loss involved. So these three of the 58 lives lost are detailed below, beginning with the remarkable man who was in the cockpit for both Eastern flights 301 and 304, first officer Grant Newby.

As recovery operations continued, reporters and CAB investigators reviewed what was known about Grant Newby and his experiences on and between both flights, a near-miss followed by a catastrophe. A compelling picture of Newby's life emerged from the official record, not only of the November 1963–February 1964 period, but also from long before he was hired by Eastern or even flew his first flight. That picture deserves a closer look

The son of a dentist, Grant Robert Newby was born March 21, 1924. He graduated from Platteville High School in 1941 and applied for entrance to the state teachers' college (the state's oldest such institution and now known as the University of Wisconsin–Platteville).[13] For the next five years, his name would appear repeatedly in the state's capital city newspapers, the *Madison Capital Times* and the *Wisconsin State Journal*, charting his very busy post–high school and war experiences.

The entrance process for the state teachers' college included freshman entrance examinations. On October 22, 1941, H.C. Wilkerson, serving as the tests and measurements instructor at the college, proudly announced to the newspapers that 26 of those examined at Platteville were ranked in the "highest 10 per cent of over 5,000 students from 27 teachers' colleges of the United States." Newby scored in the "90 to 95 per cent" range on the test and was admitted.[14] His first year he seemed to live up to the promise of his entrance exam; he was named an honor student at the end of his freshman year.[15]

The outside world entered many American backwaters and upset carefully laid plans for Newby and his generation in the fall of 1941. With Hitler approaching Moscow and the Japanese polishing final plans for launching war against the United States throughout the Pacific region, Platteville keenly felt the seriousness of what was ahead but remained largely focused on such events as the Farmers' Short Course, the Pioneer Players Play, the Forensic Banquet, an Educational Symposium, the YWCA breakfast, the Athenaeum Reunion and the All-School Parents' Day.[16]

Every year in October, college and alumni national guardsmen left town for a year of training at Camp Beauregard, Louisiana. On October 21, 1940, twenty-seven guardsmen headed to camp.[17] More joined them the next October. Grant Newby kept to his studies, making each semester's honor roll, even as he ticked the time away until his 18th birthday March 21, 1942. He celebrated and made plans to join a branch of the service that perhaps offered a more comfortable billet in the war than slogging in the deserts of North Africa or over the beaches of the South Pacific. It was the Army Air Forces and the opportunity to both fly and become an officer that he was interested in. Finally, in August 1942, his enlistment in the AAF was finalized.[18]

And then, he (and many others) waited for their call to action. From August until the following February, Newby kept attending classes, and like many others in his situation must have impatiently awaited his turn to get involved. On February 15, 1943, the *Wisconsin State Journal*, in its coverage of local men and women in war service, announced that Newby had left Platteville for Nashville, Tennessee, and his initial flight training as an Army Air Force cadet.[19]

The serious war preparation was finally beginning. Grant Newby was in Nashville for basic training until the first week of April. Almost before the smoke cleared from Pearl Harbor, Hickam Field, Singapore and Bataan, the folks back home began to read in their hometown papers about the progress the many Wisconsinites like Newby made as they were trained, kitted, transported and then put under fire. For the duration and beyond, newspapers back home ran press releases from the War Department detailing just how the local boys were making out. Newby's war career is therefore fairly easy to follow, thanks to those press releases.

On April 8, 1943, Newby finished his basic training duties in Nashville and was sent down to Maxwell Field (now Maxwell Air Force Base) in Montgomery, Alabama. He was now officially enrolled as an aviation cadet in the Army Air Forces pre-flight school for aspiring pilots.[20] The *Wisconsin State Journal* carried the proud news into every home in the Madison capital region, as well as Platteville itself, thus beginning over two and a half years of coverage, not only of Newby, but of all the local boys involved in the war. Preflight school and officer corps training occupied the next six months. In November, Newby was reported by the War Department as graduating from a new army air training field near Columbus, Mississippi. He "received the silver wings of a flying officer and the commission of a second lieutenant."[21]

Newby's experience (bouncing around the country within the chaos of the initial induction and training of huge numbers of men and women for war service in the early months) was typical. He started in Nashville, did some first flight training at Montgomery, and also gained experience at flying schools in Tuscaloosa, Alabama, and Greenwood, Mississippi, prior to his graduation and commissioning at Columbus. From April to November of 1943, he had been in at least five different training centers and flight schools before getting his wings. But he was successful: he had pilot's wings and was now to be addressed as Second Lieutenant Grant Newby.

The next month brought a bit of drama into Second Lt. Newby's life. Hometown papers, due to ink and paper limitations and rationing due to the war, began to be briefer in their reports and those "Our Boys in the Service" articles grew smaller and smaller. On December 8, 1943, word was sent by the War Department to Dr. and Mrs. Grant Newby of Platteville that their son, the second lieutenant, required surgery of some sort.[22] This was performed at an unspecified army hospital in Texas. The brevity of the report undoubtedly sent his parents' anxiety levels higher, and the operation was probably discussed in the usual small-town way. The details of the operation are not in his currently available records. But there was really no apparent reason for much concern. For the next six months Newby continued to train for service in Europe. Learning to fly Boeing's large and deadly B-17 Flying Fortress heavy bomber took time, but by April 1944, he was ready to go. Once in Europe, Newby joined the Eighth Air Forces' 509th Squadron of the 351st Bombing Group (Heavy), and served in it for four months, from German chancellor Adolf Hitler's birthday on April 20 until Newby's last mission on August 30. It was a very intense four months, with a lifetime of excitement packed into it.

Documents and reports are collected and maintained in a database by the 351st.org history group, which is dedicated to preserving history and documents detailing the squadron's war service in Europe. The original 351st Bomb Group (Heavy) was formed by the Army Air Service on November 24, 1942, at Spokane, Washington. It was sent home after the war in Europe ended and was inactivated at Sioux Falls Army Air Field on June 10, 1945.[23] Most of the time it was in existence the group operated out of a field at Polebrook, England, north of London and near Peterborough and Cambridge. During the three years at Polebrook, a total of 279 B-17 Flying Fortresses flew 9,075 sorties; 7,945 of those sorties dropped 20,778 tons of bombs on targets ranging from the Normandy beaches, to all the way across France and Germany and into East Prussia. The history organization says, "The gunners in the Group fired off 2,776,028 rounds of ammunition and were credited with destroying 303 enemy aircraft." Out of 311 credited missions, 124 B-17s were lost in combat.[24]

As mentioned, Second Lt. Grant Newby joined this outfit April 20, 1944, not long after it arrived at Polebrook. That his first mission was to occur on Hitler's birthday probably provided extra inducement to get in the copilot's seat of his assigned B-17, serial number 42–97144, for his first real mission. But the target was not to be the Fuehrer's birthday observances and his review of troops marching under the Brandenburger Tor in Berlin. In fact, Newby's first target was La Glacerie, France, just south of Cherbourg, and behind the Normandy beaches which were invaded less than two months later. Newby and his crew successfully dropped their payload on enemy installations such as artillery emplacements, field headquarters and troop convoys, and returned uneventfully to Polebrook.[25] Things got more serious four days later when the assignment was Newby's first trip over Germany. An air equipment depot at Erding, Germany, northeast of the birthplace of National Socialism, Munich, was the target.[26]

After the initial baptism of fire, throughout the rest of April, May and June, Newby and his squadron mates had the excitement and flight time ramped up considerably. There were raids all over Germany, including Brunswick, Sarau, Leipzig, Schweinfurt, Merseburg, Saarbrucken, Nienburg, and Schkeuditz, as well as repeated runs to the biggest and deadliest target of all: Berlin. Targets in occupied France were hit as well, especially after the D-Day invasions on June 6.[27]

The War Department was at first fairly quiet about Newby's experiences. He had downtime between raids and undoubtedly, like every other 20-year-old pilot far from home, he wrote letters back to Platteville. The War Department finally sent a press release back to Wisconsin newspapers, dated July 6, 1944. And it was big news: "Polebrook, England. Second Lieutenant Grant R. Newby, 20, son of Dr. and Mrs. G.W. Newby, Platteville, recently received the Oak Leaf cluster to his Air medal for 'meritorious achievement' at ceremonies at this B-17 Flying Fortress base."[28]

While the War Department release was somewhat short of details, it did mention that Newby had flown in 11 missions up to that time. Newby's missions number 10 and 11, on May 25 and May 30, 1944, were strikes against the eastern edge of the city of Berlin (heavily industrialized and armed and the area in which the Russian allies expected to enter the capital and finish the war), and on a Focke-Wulf aircraft assembly plant in Sarau, respectively.[29] Ten days later, another release was added to the award announcement: Newby was promoted from copilot to pilot on July 6 after he completed his 17th combat mission. His first mission as pilot in command was no easy milk run; the squadron attacked the government center of Stuttgart.[30]

With personnel shortages constant, Newby spent the rest of July as copilot and even filled in as a tail gunner for runs over Leipzig and Merseburg.[31] But his final nine missions saw him firmly in charge in the cockpit. His last hurrah in the air war came in August 1944 and was cause for double celebrations: Col. Eugene A. Romig, commanding officer at Polebrook, announced on August 18 that Newby was promoted to first lieutenant. At this point near the end of his tour of duty, the War Department noted Lt. Newby's accomplishments in such a short time in England: "Lieut. Newby, 20, is a Fortress pilot. He has participated in 23 aerial attacks on Nazi Europe, and for 'meritorious achievement' on these attacks, has been decorated with the Air medal with three Oak Leaf clusters. He is a veteran of four trips to Berlin and two to Hamburg and has flown to attacks on Schweinfurt, Merseburg, Erding, Oschersleben, Stuttgart, and other important enemy installations in Germany and France," was the text of the War Department release. Newby would end up flying a total of 34 official missions, and spend several months ferrying new aircraft into the European Theater of Operations before being ordered home.[32] In roughly a year and a half, an honor student and would-be teacher left a quiet, small, college town in Wisconsin, learned to fly heavy bomber aircraft, flew 34 dangerous missions over enemy territory, was promoted twice, and lived to tell the tale—all while he was 19 and 20 years old.

As the Soviets tightened the ring around the Berlin capital region and the Japanese began preparing for invasions of their home islands in the spring of 1945, First Lt. Grant Newby returned to Platteville for a well-deserved leave. He began visiting his parents on March 9. His orders were to report to a California air base for further assignment when his home leave expired.[33]

For the next year, the papers were silent about 1Lt. Newby's war experiences. He resurfaced on May 27, 1946, Memorial Day. In Lancaster, Wisconsin, the Boy Scouts planted trees around the town's Civil War monument as a way to honor the town's three World War II casualties: "John Day, pilot killed in an airplane crash; Charles Hoskins, who fell in the battle of the Bulge; and David Schreiner, killed on Okinawa." Newby was asked to give the Memorial Day address at the Grantland Theater after the tree planting ceremonies.[34]

War and memoriam over, it was time for the 22-year-old Platteville student to decide what to do next. There's an old joke among aviators: "After you've flown with eagles, it's hard to stay on the ground with turkeys."[35] It took a few more years for everything to fall into place, but he eventually signed on to Eastern Air Lines as a first officer and began to learn how to fly Eddie Rickenbacker's Golden Falcon and Great Silver Fleet aircraft, mostly Martin 2–0–2 and Martin 4–0–4 twin engined, propeller-driven planes, with space for 42 passengers on the 2–0–2 and 40 on the 4–0–4. Newby built up time and experience and was chosen as a first officer for Eastern's brand new Jet Age airliner: the Douglas DC-8.

First Officer Grant Newby had an unknown but significant number of miles on the B-17, as well as almost two million miles and 10,700 hours of flight time on a variety of airliners, as well as an incredible range of life experiences when he piloted Eastern 304 into the early morning darkness of February 25, 1964.[36]

While Grant Newby fought the Germans from above, an effort which included bombing French targets from the air, Marie-Hélène Lefaucheux fought on the ground. Her biographical information, available from French Senate and National Assembly sources, reveals an incredible wartime career.

Marie-Hélène Lefaucheux, née Postel-Vinay, was born in Paris on February 26, 1904.[37] Her death on Eastern Flight 304 came just a day before her sixtieth birthday. That sixty years was spent fighting, first for the restoration of her nation and then for the rights of women and children in all nations.

She was the second woman ever to be admitted to the prestigious Ecole des Sciences Politiques (School of Political Science) in Paris, but she left before completing her degree in order to marry Pierre Lefaucheux, an engineer, in 1925. Pierre volunteered for the French army in 1917 and received the Croix de Guerre (War Cross of 1914–1918) with two citations. He worked for a manufacturing company after the 1918 armistice. The couple's orderly life, like so many others, came to an end in 1939. After the Germans began Fall Weiss (Case White), the invasion of Poland, on September 1, Pierre was ordered to rejoin the reserve force. Since he was 42 years old at this point, he was not ordered to active duty, and was instead appointed as a director of a Le Mans ammunition factory.[38]

After the Germans ended the Phony War in May/June 1940 with Fall Gelb (Case Yellow—the invasion of the Low Countries) and Fall Rot (Case Red—the invasion of France), Marie-Hélène's family was immediately caught up in the war. Her brother Roger died in the initial German invasion; a second brother, André, was captured but escaped to England. Marie-Hélène and Pierre immediately became part of the French resistance. Pierre and André Postel-Vinay would eventually coordinate the movement of escapees and Allied pilots to freedom in England.

Marie-Hélène was appointed to the Comité de Libération (Committee of Liberation) in Paris, while Pierre became a division commander and coordinator of the almost 30,000 French Forces of the Interior (Forces Françaises de l'Intérieur—Charles de Gaulle's formal name for underground forces in German-occupied territory).

On June 3, 1944, the German SD (officially the Sicherheitsdienst des Reichsführers-SS), the intelligence agency of the SS and the National Socialist Party, arrested Pierre. The SD interrogated him at Fresnes, then sent him to the Buchenwald concentration camp in Germany on August 15. Marie-Hélène showed her courage by attempting to follow her husband, who was put on a bus with other prisoners, but she lost the trail and was forced to return to Paris for a short time. After the liberation of the capital, she headed east again after learning, through the underground network, where her husband was taken. Before the Germans retreated, Mme. Lefaucheux managed to convince or perhaps bribe a Gestapo officer to release Pierre during the chaos of the German collapse in the west. The couple managed to avoid the dangers of the front lines and return to Paris.

For their five years of wartime work in occupied France, both Pierre and Marie-Hélène were awarded the Croix de Guerre and Légion d'Honneur; Pierre also received the Croix de la Libération, and Marie-Hélène the Médaille de la Résistance. With the war over, the focus shifted to rebuilding France in the new Fourth Republic. Mme. Lefaucheux remained in political circles, but Pierre returned to industry. The French government nationalized the manufacturing concern of Renault and Pierre was appointed its CEO, a post he would hold for the next ten years until his death in an auto accident near Saint-Dizier, east of Paris.

Mme. Lefaucheux began her postwar political career as a member of the first Constituent Assembly, representing the Department of the Aisne; she also was vice-president of the Municipal Council of Paris. The assembly elected her as a counsel for the republic between 1946 and 1947, serving on the Foreign Affairs committee and the Committee of France Overseas. She specialized in issues surrounding the French Union, which had charge of colonial and overseas territorial issues until 1958. Thus, she became a member of the Assembly of the French Union in 1947, and served as its vice president in 1950. Her internal French political activity culminated in an appointment to the Economic and Social Council between 1959 and 1962.[39]

Mme. Lefaucheux during this period was also focused on larger issues of international concern. She was a member of the French delegation to the General Assembly of the United Nations and was subsequently named chairwoman of the United Nations Commission on the Status of Women. The commission was chartered to "win full political rights for women all over the world." It was a tall order; as reporters noted in 1947, as even "progressive" Belgium had granted women the right to vote only in 1947.[40]

Still, she pressed on. UN records show that, in a 1951 meeting of the commission at Lake Success, New York, Mme. Lefaucheux used her chair to highlight "the plight of a small number of women in her country who were survivors of human experimentation by Nazi doctors in concentration camps during World War II." The UN Economic and Social Council took up the issue and began negotiations with the government of the Federal Republic of Germany (the nation formed out of the French, U.S., and British occupation zones in the west). The FRG took steps to open the way for compensation for the victims, although many of them were not located. Many of them were "injected with malaria, sterilized, subjected to freezing and high altitude experimentation," and many died before help came their way.[41]

In 1957, Mme. Lefaucheux became chair of the International Council of Women. In June and July of 1963, the ICW met at Washington, D.C., to observe its 75th anniversary. She once again used her chair to highlight a critical issue. She told the attendees: "Nowhere do the mother and father have equal rights over their children except in four Scandinavian countries and West Germany." In France, a "mother cannot get a passport to take a child out of the country without the father's authorization. In Africa, if the father dies, his family takes over the children, the mother has no rights." She concluded her remarks: "A home in which harmony prevails is one where parents work out between them what is best for the children."[42]

In February of 1964, Mme. Lefaucheux traveled to Mexico City for yet another women's conference. She was returning to Paris, with a stop in New York. She boarded Eastern Flight 304 in Mexico City.[43] Her body would not be recovered from Lake Pontchartrain for 23 days.[44] It was a tragic end to a life lived in struggle to make the world a better place with more justice and equality for everyone.

There was a coda, of sorts, to Mme. Lefaucheux's story. A year after her death, Larry Collins and Dominique Lapierre published their 1965 book, *Is Paris Burning?*[45] The book was about the last moments of the German occupation, the actions of the Resistance and Adolf Hitler's orders to burn the city to the ground before the Allies arrived. The book was in turn made into a major film, with full Hollywood treatment, in 1966. The film's screenplay was penned by Gore Vidal and Francis Ford Coppola. Early in the film, Leslie Caron portrays the wife of an important political prisoner being sent to Buchenwald by the Gestapo. Her scenes concern her frantic efforts to get her husband released. This bit ends with Caron's character at a railroad station, watching her husband get shot. Caron does not appear again in the rather lengthy film. The writers acknowledged that the Caron part was "paraphrased" on Mme. Lefaucheux's own adventures (which arguably would have made a better picture in and of themselves). As truncated as it is, it does give a glimpse into the character of the real Marie-Hélène Lefaucheux and how much the world lost by her death.[46] Mme. Lefaucheux's body was recovered from Lake Pontchartrain on March 19, 1964. She is buried with her husband in Saint-Quentin-des-Prés.[47]

In the spring of 1939, an Iowa high school senior named Basil Rowland had the responsibility of penning a "Bits of Nothing" portion of the student news page in the *Fayette County Leader*. For his March 23 entry, Rowland noted a repeat visit by a bass-baritone singer named Kenneth Lee Spencer.[48] "Not only an outstanding singer and an excellent actor, but a very pleasing personality—that's Kenneth Spencer. He always seems to be in good humor, even after long hours of rehearsal and concert. You may expect to hear of Kenneth Spencer again.... We are confident that he is going places," Rowland wrote.[49]

Fayette, Iowa, a small town about 150 miles northeast of Des Moines, with a 1940 population of about 1,100, is the home of Upper Iowa University, which hosted Spencer's March 16 performance. The university's choir was given the profits of the recital, which cost the general public 35 cents and students 25 cents.[50] Spencer's recital was so well-received in Fayette that he played a return engagement there in May 1940 after completing a run of *John Henry* in Philadelphia, New York and Boston. He made several stops in Iowa on his way to a month of appearances in Winnipeg, Canada.[51] The type of enthusiasm that met Spencer all over Iowa was not new, nor would it be the end. He was a remarkable singer and actor whose life also ended on Eastern Flight 304.

Born the son of a steelworker on April 25, 1913, in Los Angeles, Kenneth Spencer worked odd jobs during the hard years of the 1930s, taking private vocal lessons whenever possible. His deep, rich bass-baritone voice, quick study of languages and commanding presence helped him gain a scholarship to the Eastman School of Music at the University of Rochester in New York.

Following graduation from that school in June 1938, he joined the St. Louis Civic Opera Company to sing the lead in the touring production of *Showboat*. A Hollywood Bowl performance as well as the touring and stage experience brought him to the attention of motion picture executives. He debuted in supporting roles in two signature films of 1943, MGM's *Cabin in the Sky* and *Bataan*.[52]

The musical *Cabin in the Sky*, the film version of a Broadway hit, was released in March of 1943. The story is a retelling of Goethe's Faust legend: a man killed over gambling debts is given six months to prove himself worthy of heaven or he will be consigned to hell.

The movie starred a "who's who" of African American actors and singers, from the leads of Ethel Waters and Eddie "Rochester" Anderson to Louis Armstrong, Rex Ingram and Lena Horne. Spencer actually portrayed two parts: Rev. Green and the General, the commander of heavenly forces fighting for the soul of gambler Little Joe Jackson (Anderson).

The all-black cast was directed by a white man, Vincente Minnelli, in his first directing job. Minnelli, Judy Garland's second husband and father of Liza Minnelli, found success artistically with the film, going on to direct many classic American films, such as *Meet Me in St. Louis* (with Garland), *An American in Paris* (with Gene Kelly) and *Father of the Bride* (with Spencer Tracy). Seven of his actors garnered Academy Award nominations.

The stars took some risk to make the picture, as did Minnelli and MGM. Many theaters, especially in the South, would refuse to show it. But receipts proved them wrong; the film cost somewhere around $600,000 and grossed $1.6 million.[53]

Three months after his strong voice helped him stand out in his screen debut, Spencer appeared in a more timely, grittier film, *Bataan*. This film was no fantasy; it was based on the all too real (and very recent) experience of soldiers on the Bataan peninsula just a year earlier as the Japanese advanced through the Philippines. The actual battle famously ended in the Bataan Death March, in which between 60,000 and 80,000 Allied soldiers were force-marched under brutal conditions to POW camps.[54] In the film version, 13 Filipino and American soldiers are left behind at a key spot and ordered to destroy a bridge (and prevent any rebuilding attempts) to slow down the Japanese advance on Manila.

Bataan starred Robert Taylor, George Murphy, Thomas Mitchell, Lloyd

Nolan, Lee Bowman, Robert Walker, and Roque Espiritu, with Spencer portraying engineer Private Wesley Epps. The cast also included Desi Arnaz, who was soon to marry Lucille Ball and gain everlasting fame as Ricky Ricardo.[55]

Lena Horne, in a 1943 interview about *Cabin in the Sky*, said, "All we ask is that the Negro be portrayed as a normal person." Spencer's performance in *Bataan* may well have been much closer to what Horne meant than the fanciful Minnelli musical. *Bataan* was one of the first Hollywood films to portray a black character as something other than a minstrel or performer, as just a normal human being. And that portrayal was all Kenneth Spencer.[56] MGM production chief Dore Schary later claimed he cast Spencer in the role to "break the color barrier in American war films." There were actually no black combat soldiers in 1943; black members of the U.S. armed forces performed strictly support roles, mainly as menial labor such as dock workers and cooks. The first black soldiers would not arrive in combat until July 1944, when the 370th Regimental Combat Team went ashore at Naples to help in the Italian campaign. Schary would add that MGM received many letters of complaint about the Wesley Epps role. But Spencer's portrayal of Epps was a sea-change and brought awards, including from the NAACP.[57]

Roles for black actors, even successful ones, on Broadway or in Hollywood were very limited. In 1946, Spencer starred in another revival of *Showboat*, to critical acclaim and a one-year run. He continued singing on the recital circuit and on radio. But he was reportedly growing increasingly frustrated at the difficulties of performing in the U.S., much less the nonavailability of stage and film roles. How many times could he stand singing "Ol' Man River" and how many revivals of *Showboat* would come along? In 1949, he gave his first European performance, in France at the International Music Festival, and it proved to be a an eye-opener for Spencer.

Offers and accolades from that performance came in from all over the continent, especially Germany and France. Encouraged, Spencer moved to Germany in 1950. He worked in a variety of concert, film, musical, opera, radio and stage productions and released a number of performances on the Columbia Masterworks label.[58] A few of his performances are available today on various Internet sites; it can be startling to hear him singing in German, but there is no denying the incredible richness of his voice.

In February of 1964, Spencer was in New Orleans for a concert, one of the first since he left the U.S. Things had improved, if only incrementally. When he boarded Eastern Flight 304 at Moisant International, Spencer was able to use the general airport facilities and waiting areas only because segregation had just ended at the airport and its leased restaurants and bars only recently.

In 1961, the U.S. Justice Department sued airports across the South for segregating black passengers in small, "coloreds-only" waiting rooms with minimal, separate water fountains and restrooms. New Orleans, and Eastern 304's next destination, Atlanta, were among those southern airports that finally integrated after losing the lawsuits, which were initiated partly at the request of the CAB and FAA. Individual lease holders integrated in the months following. In fact, one restaurant operator at Moisant, Interstate Host, Inc., lost its own class action lawsuit and ended the practice of prohibiting colored passengers from dining in its "International Room" on August 8, 1963, just six months before the Eastern crash.[59]

Spencer's body was never recovered from the deep crater in the mud of Lake Pontchartrain.[60] He was 50 years old.

10

"Extreme disintegration of the aircraft structure": The Investigation of Eastern Air Lines Flight 304

The extreme disintegration of the aircraft structure precluded any crash/injury study. The crash was non-survivable.
—CAB Aircraft Accident Report, July 1, 1966

Multiple Locations
1964–66

The investigation into the loss of Eastern Flight 304 began almost from the moment the aircraft disappeared from radar scopes. As the night ended and daylight broke, the aircraft didn't show up with inoperative radios at its intended or unintended airport, nor was news received of any crashes between New Orleans and Atlanta. Since officials knew that the aircraft's last reported radar position was out over Lake Pontchartrain, they also knew the whole situation they were facing was almost exponentially more difficult, even though the lake was only 16 feet deep.[1]

Once a possible site of impact was determined, obviously the next step (beyond the imperative of looking for survivors and recovering the bodies of the dead) was to bring up as much of the aircraft as possible, even in the face of all difficulties. Conditions were somewhat reminiscent of the Northwest Flight 705 site: mud, snakes and an airplane deeply embedded in a crater. But this time, the 20-foot-deep crater was also underwater—and filled with viscous mud. Finally, what parts were not in the crater or along the lake bed floated in a wide area, pulled by tides to the Gulf of Mexico.

Two 24-hour-a-day dragging operations began at the site of an oil slick floating on the water and at the point where radar contact was lost. Yet even this thorough and exhaustive effort was not immediately successful. The exact spot where the majority of the wreckage was embedded was not discovered until late in the day on March 13. Salvage of what was recoverable at this point continued nonstop until April 16. Authorities estimated that just 60 percent (by weight) of the aircraft was recovered by this point, a month and a half after the crash.[2]

One of the pieces highly sought after was the flight data recorder. Unfortunately, it was not very helpful. FDRs have been around in one form or another since the World War II years, if not before. In the case of Eastern 304, the Fairchild Model 5424 FDR was only partly recovered from the muck below. The recorder magazine, record spool and some 50 feet of loose and unused tape was rescued. But "the last readable portion of the tape was 150 minutes of flight, encoded as Flight 304 of the 24th, ending at a point which appears to be the landing approach to New Orleans. The takeoff portion of the tape was not recovered." The FDR was silent on what the DC-8 experienced on the fatal flight segment.[3]

There was no cockpit voice recorder aboard N8607; until July 1966, CVR systems were not required on the airlines' fleets, and the airlines were still at this point fighting the possible mandate through heavy lobbying of Congress. Given the condition of the FDR, having the cockpit conversation would have been crucial to finding out what happened. In fact, this was to be exactly what would happen two and a half years later in Falls City with Braniff Flight 250, as will be detailed in later chapters.

Even as workers—especially expert divers, who could barely see a foot in front of their masks—continued their efforts, often interrupted by rain, mist and fog, CAB officials suspected they had yet another jet-upset-in-turbulence investigation on their hands, this one the deadliest yet.

As with Northwest Flight 705, Eastern 304's wreckage was taken (by barge instead of helicopter) to a hangar at a nearby airport. In this case, the smaller New Orleans Lakefront Airport, which was closer to the crash site, was utilized. All parts were examined and detailed notes on condition, position and location (among other categories) were made. They were washed and then laid out in the airport hangar to be pieced back together.[4]

The CAB reported that they recovered at least portions of "all extremities" of the DC-8 from the main impact spot. These showed a high degree of fragmentation; the largest piece left was just five feet of the upper portion of the rudder. Flaps and landing gear proved to be in the up and stowed positions.[5] And more important, the recovered pieces showed that there was no fire, explosion or structural failure in midair. The aircraft fragmented when it hit the water, not beforehand. Lamb told a press conference that impact fragmentation was probably why the jet's large tail section wasn't sticking up out of the lake. It, too, was in small pieces.[6]

All four engines were eventually extracted from the lake and minutely studied. No evidence was found in any of them of "operating distress" prior to impact. The four Pratt and Whitney JT4A-9 engines had been used between 6,300 and 7,800 hours, while the airframe itself had 11,340 hours—all in the low-average range. The numbers one and two engines were discovered within 45 feet of numbers three and four, meaning none of them had separated in flight. All four powerplants had similar severe disintegration impact damage, but it was possible to tell that the fuel system was operating normally.[7]

One particularly telling finding where the engines were concerned was that the recovered reverser assemblies showed reverse thrust was being used at impact.[8] The crew obviously attempted to slow the falling aircraft; it was not lost on investigators that this technique partly helped Eastern Flight 301 recover over Texas, and that the pilot flying the aircraft in both cases was First Officer Grant Newby. The technique used by Captain French and Newby saved the Flight 301 aircraft, but caused an engine to be ripped away. In Flight 304's case, the engines stayed on; but the plane was flying at a lower altitude and ran out of airspace before recovery could be made, and it impacted the lake.[9] This was a significant finding, but while the investigators knew why Flight 301 was upset, they did not know what caused Flight 304's dive.

The board looked at the aircraft's maintenance log in its early days, since the log was available from Eastern at the airport. It showed five entries of items which were to be corrected when N8607 arrived in New York: "1. Fuel totalizer reading wrong, 2. Outer pane center windshield heat inoperative, 3. Number three engine ejector light blinks, 4. Number 3 main fuel gauge reads 2–4,000 pounds high." Item 5 on the log book was the inoperative pitch trim compensator (PTC).[10]

An airliner is often dispatched with less than perfect components. There is a "Minimum Equipment List" (MEL) for aircraft, mandated by government regulators or the airline itself. If any of the MEL items are inoperative, the plane does not leave the ground.[11] None of the five items on N8607's log were necessary for safe operation or on the MEL. Some of them were annoying (the blinking engine ejector light), while there were ways to compensate for others (the number three fuel gauge inaccuracy). Still, all five of these areas would receive CAB scrutiny, but since the PTC was the only one of the five which was part of the flight control system, it would undergo special attention. And since there were weather reports of turbulence in the area of a severity to match the conditions encountered by Flight 301 three months earlier, there would be similar scrutiny from the CAB team assigned to weather factors.

10. "Extreme disintegration of the aircraft structure" 85

In its final report, 30 pages long, the CAB went to great lengths to explain how an aircraft, particularly the DC-8, is controlled longitudinally. For us nonpilots, it helps to stop a moment and review the basics of flight control and just what "longitudinal control" means. Drivers know that you use the steering wheel to turn left and right on the vertical axis (which could be thought of as a pin driven vertically through the roof of the car to the pavement and located at the center of the vehicle). This left/right movement in a car is called "yaw" in an aircraft, and is produced by the rudder pedals on the floor of the cockpit. Fairly simple: you push forward against the rudder pedals with your left leg to produce a leftward yaw, and with your right leg to produce a rightward yaw.[12] One of the first difficulties an experienced driver has in taking flying lessons (this writer included) is learning that yaw is controlled by your feet and not your hands and arms.

Where a car has this one axis of movement, an aircraft has three. Besides yaw along the vertical axis, there is movement along the lateral axis (which runs side-to-side from the left wingtip to the right wingtip, again through the center of gravity). Movement around this axis is called "pitch"—the nose "pitches" up and down. When the nose is up, the tail is down, and vice-versa.[13] Where the steering wheel of a car is fixed and does not move back and forth, the steering column of an airliner like the DC-8 can be pushed or pulled. Some airliners, particularly those of the Airbus consortium, use joysticks mounted on the side console of the cockpit, but the principle is the same. When a pilot pulls back on the steering column, or joystick, the aircraft nose pitches up. When the column, or joystick, is pushed forward, the nose pitches down.[14] And unless you drive your car off a cliff or try to drive it up a steep mountain, a driver does not have to control pitch. The last axis of control on an aircraft is the longitudinal, a line running from the nose to the tail. Rotation around this axis is called "roll" and is controlled by the wheel mounted on the column, or joystick, when it is moved left or right.[15] In some cases, a car can be rolled, but it usually ends badly for everyone concerned and is usually not experienced by the average driver.

An automobile's control mechanism does have one commonality with an aircraft: when you open the throttle by pressing on the floor-board accelerator, the engine produces more power and goes faster; let up on the accelerator and the car slows down. An airliner has the same system, except the throttles for multiengine aircraft are usually mounted in the center console between the two pilots, one throttle for each engine. When a pilot pushes the throttle forward towards the dashboard, the engines produce more power. When he pulls them backwards, the engines slow down towards idle. First-time flying students have to adjust to floor pedals producing left-right movement, instead of speed. (There are at least brakes for an airliner in somewhat the same spot as on a car; they are located at the top of the rudder pedals.)

Non-pilots often don't understand the physical coordination required to actually fly an aircraft, from a one-place, single-engine trainer, up to the mammoth Airbus A-380. In an automobile, coordination of speed using the right foot for acceleration and left foot for braking, plus turning the steering wheel left or right to turn the car, is about the extent of the basic driving requirements. Up in the air, however, all three axes, plus engine power, aren't controlled separately, one at a time; turning left can involve pushing in the left rudder pedal, turning the steering wheel, or yoke, to the left, and applying some pressure to hold the control column somewhat back towards the pilot, in order to keep the nose up during the turn.

This kind of coordination is learned fairly quickly and becomes second nature as student pilots get more and more air time. And this coordination is almost never the only thing going on in the cockpit. There are radio messages to listen to, understand and reply to, conversations with copilots and, in some cases, flight engineers. A bewildering array of gauges and instruments (and how these instruments are designed and placed would figure in to the Eastern 304 investigation) require attention; and navigation under a strict set of rules, controls and radioed instructions has to be figured out and followed. Add bad weather, turbulence, darkness, disorientation, physical or mental exhaustion or distractions, problems with passengers, and on and on and on, and it becomes somewhat miraculous that the airline, general and military aviation systems and aircraft function at all.

Over the development years of the airline industry, systems within the aircraft became more and more sophisticated to control more and more aspects of flight. Not only was this necessary for flight control, it was necessary for safety. But the days of simply starting up an engine, pushing the throttle forward and controlling the three axes of flight by a stick and two rudder pedals were long gone by the time jetliners took to the sky.

The CAB decided to devote a sizable portion of its final report to discussing how the Douglas jetliner was controlled. Because first appearances were that Eastern Flight 304 had fallen from altitude into the lake (i.e., pitched down into impact), a look at the longitudinal controls of the DC-8 was needed.[16] Again, all of this was foreshadowing the direction the board would take to arrive at its probable cause statement. The question was, they felt, whether the aircraft's longitudinal system failed because of turbulence/jet upset or whether it failed on its own. Understanding how that system works normally was a key to the process.

Understanding the three axes of flight and the pilot inputs which control them is a start. But what happens exactly between the pilot's control input and the part of the airplane which causes change along those axes? In this case, the CAB explained how Douglas made it happen on the DC-8:

> The aircraft can be controlled longitudinally by use of the elevators or variable incidence horizontal stabilizer."[17] [The horizontal stabilizers are the two smaller wing-like structures located usually at the rear of the aircraft on top of or underneath the rudder. The elevators are attached to the rear of the stabilizers. The entirety of the stabilizers can move up and down along a lateral axis of their own; the elevator portion can move further up or down for additional control.] The elevators are operated by movement of either control column through two independent cable systems to elevator control tabs. The elevators are connected together by a torque tube at the rear spar of the horizontal stabilizer.... Tabs on the trailing edge provide aerodynamic boost to control inputs.[18]

This complex system is linked to the pilots' controls in the cockpit. And pilots have a feel for what is happening with control surfaces they can't see by wheels, pulleys and springs in tension. The force a pilot uses to get these control surfaces to move is called "stick force," which is, according to the CAB, "the most often used of several terms, including 'column force' and 'wheel force,' to describe the pull and push forces required of the pilot to operate the elevator control whether it be a stick, column and wheel, or shaft and wheel." The report continued: "Most of the pilot's stick force is provided by a load feel mechanism with two opposing preloaded springs which establish a neutral point of the elevator control system. Resistance of these springs to motion of the control column is greatest near the neutral point."[19]

The report now got less basic and closer to the point. The design of high-speed jets often produces a nose-down pitch movement, said the board. So Douglas designed the "Pitch Trim Compensator (PTC) system" to counteract this natural aerodynamic force. The PTC applies nose-up control through the elevator system. Ah, finally we non-pilots have some understanding: an aircraft pitches nose up or nose down and this pitch is controlled by pilots moving controls forward or back, which makes the stabilizer/elevator combination move to a position which moves the nose up or down. Adjustments of these controls help "trim" (adjust) the nose up or down during various portions of a flight, as external forces act on the aircraft's aerodynamic properties.[20]

The board now returned to the DC-8's PTC. The board noted that "operation of the PTC is also required in the low altitude, high speed regime below mach effect to improve stick force characteristics as speed increases. This system consists of an electrical computer, an electrical actuator, spring loaded linkages, and a mechanical indicator." The computer senses whether the aircraft is at high altitude or below 20,000 feet and "provides electrical signals to the actuator which actually moves the copilot's control column." There is an important distinction there at the end of that sentence. When the pitch trim system senses that the nose needs to be trimmed up or down, its actuator moves the *co*pilot's column on the right side of the cockpit, *not* the captain's on the left.[21] In the case of the two flights we are concerned with here, actions which affected pitch trim on Eastern 301 and 304 were felt in the control column on the right. On both flights, these inputs were felt by First Officer Grant Newby.

The board continued its Aircraft Control 101 lesson with further detail on the inner workings of the PTC unit itself. When the need for pitch trim begins, the PTC actuator kicks in; this begins at "either Mach 70 or 310 knots and increases in displacement and rate up to Mach 88 or 410 knots. The maximum input is 36 pounds of stick force [pressure which is felt by the copilot]. Actuation of the PTC is indicated by the extension of a plunger from a flexible cable housing attached to the left side of the copilot's control column. There is no measurable correlation between the amount of indicator showing and the degree of actuator extension." The copilot's indicator is not a gauge then, but more of an "it's on/it's off" type of indicator. Also, "a three-position switch located on the left side of the control pedestal permits normal operation, testing of the system in the spring-loaded test position, and an override position which may be used to retract the actuator in the event of a malfunction."[22] The PTC is basically all the first officer's baby.

Longitudinal trimming (the adjustment of stabilizers and elevators) from a mechanical point of view occurs by "hydraulic or electric actuation of the horizontal stabilizer. The hydraulic motor trims at a rate of a half-degree per second" and it can move the stabilizer between "10 degrees aircraft noseup (ANU) to two degrees aircraft nosedown (AND)." This motor actuates by "manipulation of dual toggle switches on either control column, or by split 'suitcase' handles mounted side by side on the center console. The electric motor trims at a rate of one-seventeenth of a degree per second, and is actuated by dual toggle switches on the center console, or by the autopilot." The system can only be actuated by "simultaneous operation of any set of dual switches or handles"[23] on the center console between the two pilots. (A failure of a jackscrew in the stabilizer system of an Alaska Airlines Douglas MD-83 off the coast of California in 2000 would result in the deaths of all 88 onboard. In that case, when the jackscrew failed the aircraft became uncontrollable along its longitudinal axis and fell into the Pacific Ocean. The NTSB determined probable cause in that case as "a loss of airplane pitch control resulting from the in-flight failure of the horizontal stabilizer trim system jackscrew assembly's acme nut threads. The thread failure was caused by excessive wear resulting from Alaska Airlines' insufficient lubrication of the jackscrew assembly."[24] Alaska Flight 261 is a good example of why aircraft have horizontal stabilizers in the first place.)

Now that we know how the pilots use which controls to control the aircraft's nose-up or nose-down needs, the report explains what happens in the belly and rear of the aircraft: "Both motors provide power through differential gearing to a drive shaft on which a dual sprocket assembly is mounted. The sprockets are connected to the common drive shaft by shear rivets, and each transmits rotation of the drive unit through roller chains to an irreversible jackscrew. Failure of either set of shear rivets freezes the stabilizer in the last selected position, and further operation is impossible." Back in the cockpit, "the indication of stabilizer position is provided by fore and aft movement of a small 'bug' along a scale on the left [captain's] side of the center console."[25] The modern jetliner can also be controlled longitudinally by using the autopilot system, which "utilizes elevator displacement to initially retain the selected attitude. An automatic trim coupler senses elevator servo torque information and generates stabilizer trim commands when torque of a given value or time interval is encountered. Any 'runaway' or contradiction in the system results in the interruption of power to the autopilot and the illumination of a warning light."[26]

Returning now to the issue of instrumentation and its visibility and placement (the importance of which was mentioned above), the board's report discussed a small instrument called the attitude indicator. In N8607, the accident aircraft, attitude information was provided by a Collins 105 Approach Horizon. The center of this gauge shows movement of the 'miniature airplane' on the black face of the instrument. The gauge had no marks for the degree of pitch. The board noted this device's limitations were enough to render it useless. "It is possible for the instrument to indicate a reduced rate of pitch when attitude changes through 20 degrees of pitch, even though the actual rate of change is constant. In a corresponding manner, if the attitude has exceeded 20 degrees, the displayed rate of aircraft response to control inputs will be slower than the actual response."[27] The gauge was slow, said the CAB, and this affected the quality of information received by the pilots.

The board's analysis of all these operational aspects would come later in the thirty pages of the final report. But first, it discussed the weather and the extent and severity of rough air north of New Orleans International. Investigators went to great lengths to study and test and find out what role turbulence could have had in this jet upset. Knowing that the previous year's deadly upset of Northwest Flight 705 (and Flight 301 and others) was brought on by turbulence, and with rough air known to be over Lake Pontchartrain, would this Eastern crash go down as the second in a series of crashes which could shake and even destroy confidence (publicly and privately) in the aircraft? The weather investigative team got right down to business.

Early on, the board's team knew that the accident aircraft, N8607, while flying an Eastern route out of Washington Dulles Airport in northern Virginia, had already been in some very nasty weather, somewhat mirroring First Officer Grant Newby's experience in the sister aircraft. Both plane and pilot dodged separate near catastrophes, but then went down together a few months later. Investigators looked into the aircraft's records while they "conducted studies pertaining to aircraft characteristics in turbulence. This information revealed that turbulence has known energies broad enough to excite aircraft natural frequencies between 0.2 cps (cycles per second) and 4.0 cps." This seems like an odd direction to go—what were they talking about? The report continued:

> An example of this is illustrated by the captain involved in the turbulence incident in N8607 at Dulles. He stated that "... we encountered the most violent jolt I have ever experienced in over 20,000 hours of flying. I felt as though an extremely severe positive, upward acceleration had triggered off a buffeting, not a pitch, that increased in frequency and magnitude as one might expect to encounter sitting on the end of a huge tuning fork that had been struck violently.
>
> "Not an instrument on any panel was readable to their full scale but appeared as white blurs against their dark background.
>
> "From that point on, it could have been 10, 20, 60 or 100 seconds, we had no idea of attitude, altitude, airspeed or heading. We were now on instruments with no visual reference and continued with severe to violent buffeting, ripping, tearing, rending crashing sounds. Briefcases, manuals, ash trays, suitcases, pencils, cigarettes, flashlights flying around like unguided missiles. It sounded and felt as if pods were leaving and the structure disintegrating.
>
> "The objects that were thrashing about the cockpit seemed to momentarily settle on the ceiling which made it impossible to trust ones [sic] senses although I had a feeling that we were inverted as my seat belt was tight and had stretched considerable. As my briefcase was on the ceiling, I looked up and through the overhead (eyebrow) window and felt that I was looking down on the top of a cloud deck. (The first officer) later said he had the same impression at the same instant as we acted in unison applying as much force as we could gather to roll aileron control to the left. The horizon bar at this time started to stabilize and showed us coming back through 90 degrees vertical to a level attitude laterally. At this time, I had my first airspeed reading decaying through 250 knots. The air smoothed out and we gently leveled off at between 1,400–1,500 feet."[28]

So, the experience aboard Eastern's DC-8 N8607 near Washington Dulles sounds similar to the experience of Eastern Flight 301 near Houston. In both cases, pilots fought and found a way to save their aircraft at the last moment. Then, on February 25, 1964, N6807, flown by the man who had been at the helm of Eastern 301, would be upset in turbulence again. Given the obvious durability of N8607 and the experience Grant Newby had recently acquired (and more than likely also built on his experience surviving 35 missions over occupied France), why was the outcome so different with EA304? Since you could fly DC-8s even while the PTC was inoperative (the accident aircraft,

N8607, and other DC-8s in Eastern's fleet were flown numerous times with inoperative PTCs which had been repaired or replaced numerous times, as noted in a long discussion on pages 16–17 of the report), it must mean that some pretty strong forces (described by the CAB as within the "man-machine-triangle")[29] were needed to bring about the demise of the aircraft and its 58 passengers.

As mentioned above, in a modern jetliner, there can be many distractions and demands on a pilot's attention. As the CAB gathered testimony and reports on the subject of speed stability, the issues of distractions and difficulties in reading instruments during turbulence and their effect on stability came up again.

In terms of the DC-8's speed stability, a test pilot with long years of experience once again took a DC-8 similar to the accident aircraft through its paces and reported: "When I trimmed the airplane at 300 knots, I found that the static stick-free stability was positive when I departed to 85 percent of trim, but when I increased to 115 percent of trim, the static stability was within the friction band and, for all practical purposes, would be called neutral." The test pilot reported that "neutral speed stability in itself does not pose a serious problem to the pilot, and, in fact, under normal flying conditions ... it is actually quite a nice airplane to fly."[30] But the test pilot had a caveat: "The thing that is dangerous about a situation like this is a distraction. If the pilot, for example, is distracted for any reason and allows the aircraft to start diverging from its trim condition, especially if he is in turbulence and he is faced with a fairly substantial change in his trim or his attitude, the *tendency usually is to make a large input, and this is where the trouble begins*"[31] (emphasis added).

This, then, was what the AMAL researchers discussed in chapter 4 discovered in their centrifuge simulations of United Flight 746: "by responding to a strong kinesthetic illusion of climb or dive after correcting from an unusual nose up or nose down attitude pilots are creating ever increasing deviations from normal flight pitch attitude in both directions alternately, somewhat analogous to pilot-induced oscillation."[32] The board's report on Eastern 304 had the AMAL report in hand, but didn't quibble about what happened being "somewhat analogous" to anything. "Overflying," or over-controlling, an airplane, rather than letting it "find its own sweet spot" at a given point while in the air is, to the board, Pilot-Induced Oscillation (PIO).[33] (The tendency over the last few years has been to move away from language such as "pilot-induced," which can imply pilot fault, towards the more neutral "aircraft-pilot coupling," or APC.[34] While such a shift away from "always blame the pilot" language is welcome, the phenomenon being described was called PIO in 1964 and that's how it will be referred to here.)

PIO can be described (and was seen in the United 746 AMAL research) as the pilot getting "out of phase" with his aircraft. For any number of reasons, the nose could pitch suddenly and unexpectedly down, which can startle the pilot and cause him to raise the nose quicker than he should, particularly if he is in the landing sequence or being tossed around the sky by a storm. The aircraft's nose then comes up ... and keeps rising, higher than the pilot wants, so he pushes down to arrest the pitch up. Because there is a pause between when the pilot acts/reacts and when the aircraft does, this quick overreaction results in wider and wider oscillations. The pilot, adrenalin flowing perhaps, finds himself asking the aircraft to pitch down, but since it is out of phase with his actions, the aircraft pitches up. This type of situation more commonly involves inexperienced and student pilots, but veterans can find themselves in similar predicaments.

How do you break the cycle of PIO? By essentially ignoring the screaming in your brain and the adrenalin pumping through the heart and by doing what seems to be counterintuitive, you stop your "large control inputs" and let the control column return to its neutral, trimmed position. Get out of the increasingly destructive loop. Since aircraft are designed to fly with natural stability, they usually return to proper trim. This assumes that there is enough airspace between the airplane and the ground. Getting caught in a PIO loop close to the ground can, and has been, fatal. In Eastern 304's case, however, the possibility that a jet upset occurred in turbulence with the pitch trim compensator inoperative, which in turn created a PIO situation, was where the CAB's thinking and analysis was heading. The test pilot testifying to the CAB underscored all of this:

> What happens is that, especially in large aircraft, we have these slow response characteristics ... as long as you fly an airplane and don't try to force it—you allow the airplane to respond well within its capabilities—you don't have any difficulty. If you, however, try to force the airplane to respond faster than it wants to, then you can get into what we call a low frequency pilot-induced oscillation. It is nothing more, really, than over-controlling. You don't see the airplane respond immediately so you have the tendency to put a little more elevator in, and by this time the airplane has started to respond and you suddenly find the response is more than you wanted. So the tendency is to reverse the process.... I can see this situation can be quite critical in turbulence or possibly under IFR conditions plus turbulence where, let's say, you do have some large gusts which change your attitude appreciably. If the pilot attempts to ... maintain his attitude tightly, there is a possibility that he can get himself involved in a PIO.[35]

With PIO a possibility, was it not precipitated, but also made worse, by turbulence producing shaking so bad that the instrumentation, already somewhat difficult to see, was made unreadable? This was, after all, what pilots in the

AMAL centrifuge study reported—that attitude indicators then in use by most airlines were not readable when the simulation got rough.

Testimony was taken by the CAB "concerning the miscues presented to pilots by their flight instruments during turbulence flying." The board reviewed research papers written in the previous few months and reported: "Generally conceded is the fact that airspeed, rate-of-climb, and altitude presentation can lack accuracy and, even more, can present completely erroneous information as to longitudinal attitude, i.e., trends exactly opposite to that expected of a given attitude. Now the Board finds that ... the attitude indicator presents to the pilot information which, while not illogical, is certainly not optimum."[36] (Notice that the CAB's language is careful. While saying that design of the attitude indicator by the Collins company and its placement on the instrument panel by Douglas were both logical, both could be better. The board did not mention that a Lear model had been more readable in the AMAL centrifuge simulation.) But Douglas did come in for some (highly qualified and gentle) criticism: "Based on the information available to the Board, the DC-8 exhibits very low speed-stability characteristics, particularly at higher climb speeds when the PTC does not operate as programmed. There was testimony at the hearing and depositions about whether these stability aspects were within the requirements of the Civil Aviation Regulations and much was made over the fact that the regulations do not address themselves to stability in the event of a mistimed condition or a system malfunction. [Under such conditions, the regulations require only that the aircraft be safely controllable]."[37] In other words, the aircraft's design met what regulations existed and was quite stable, but both that stability and the government's regulations could be better.

In terms of the people hired to fly the DC-8, the board noted, the regulations were not particularly important, but the stability was: "To the pilot the aircraft responds the same, whether or not it was required to meet any stability criteria for the condition in which he finds himself. What would be of primary interest to the pilot and is of primary interest in this report is the fact that at lower speeds (220 knots) the airplane can under certain mis-trim conditions exhibit low to neutral stick force per G and stick force versus elevator deflection, and at the higher climb speeds (310 knots) it can have very low speed-stability."[38]

Investigators therefore had three areas of interest: the PTC and the longitudinal control system (including the stabilizer drive system); PIO and the human response to a possible upset (including instrument blurring); and, above all, the turbulence which probably kicked the whole sequence into action. These three factors were not deadly by themselves: the PTC was not strictly necessary for flight; PIO and jet upsets can be reversed (as demonstrated by Grant Newby himself three months previously); and aircraft, even from the Wilbur and Orville Wright days, had flown successfully in turbulence. But when severe turbulence upset the aircraft, the combination of all three factors resulted in an almost unrecoverable situation.

Given that the dive towards the lake began at slightly less than 8,000 feet, there was little time to overcome the three problems faced by the crew. If Flight 304 in February had been as high as Flight 301 in November (almost 20,000 feet) there might have been time to effect yet another recovery. After all, the man—Grant Newby—who had been at the controls and then helped recover Flight 301 was also at the controls and helping Flight 304. The final report noted this juxtaposition as "an additional noteworthy element" but stated that it "is impossible to assess."[39] Still, they put the details into the official record:

> It is most probably, based on voice identification and crew practices, that the first officer was at the controls during and following takeoff from New Orleans. Likewise, this same pilot was at the controls during the development of a longitudinal upset in another DC-8. There are still many unanswered questions concerning the exact mechanics of that earlier incident, but it is known that this pilot did not hesitate to apply full forward control column, and additional nose-down trim, when faced with an unusual attitude in turbulence. [Based on crew testimony, the amount of trim used in that case was probably less than the 2-degree AND limit.] The result was a dive reaching about 40 degrees nosedown and from which about 13,000 feet were required for a recovery. The Board fully recognizes that what this pilot did in one situation at one time is not necessarily indicative of his actions in another, even similar, situation at another time. While the Board admits to the subjective nature of this information, it cannot ignore its existence.[40]

Investigators were right; comparing the two experiences is indeed of a "subjective nature." But, as the board found, it is hard to ignore what happened: Grant Newby needed about 13,000 feet to recover Eastern 301 from jet upset in turbulence. When the same thing happened on Eastern 304, he immediately reacted and used the same technique, including deploying the thrust reversers on the engines—but he had only 8,000 feet to spare. This time, his DC-8, although basically level horizontally, was still descending at a high rate of speed when it ran out of air space, striking the lake with enough force to displace 16 feet of water, create a 20-foot-deep crater in the mud, and splinter a $5.5 million aircraft and its human cargo into pieces.

CAB investigators began to wrap up their duties and write the final 30-page report by the summer of 1966. On page 26 of that report, the investigative team listed its official "findings" on Eastern 304 after almost two and a half years of dredging in the mud of Lake Pontchartrain, testing, testimony, analysis and discussion. There were seven findings of fact on the list:

1. Night, instrument conditions prevailed.
2. Moderate to severe turbulence was encountered.
3. The PTC was inoperative and may have been partially or fully extended.
4. The stabilizer drive system failed in the 2-degree AND [aircraft nose down] position at some time during the flight.
5. The attitude indicator, which was small with a solid black background, was difficult to interpret at night.
6. The pitch indication of the attitude indicator was "geared down" but not indexed as to degrees.
7. The aircraft exhibited marginal to nonexistent speed stability and a stick force per g characteristic which test pilots have interpreted as unstable.[41]

Therefore, the "probable cause" of Eastern Air Lines Flight 304 crashing in the early morning hours of February 25, 1964, was "the degradation of aircraft stability characteristics in turbulence, because of abnormal longitudinal trim component positions."[42]

The language of probable cause is always very deliberate. Even though CAB/NTSB findings of probable cause are not admissible in civil or criminal court (more on that in a following chapter), the results still carry heavy political and commercial implications. The agency usually always has a reason for wording statements of probable cause in specific ways. In this case, they could have altered the position of the phrases in the statement by writing "abnormal longitudinal trim component positions caused the degradation of aircraft stability characteristics when it was flown into turbulence." "Abnormal positions" and "degradation of characteristics" are fairly euphemistic. The board was surely confident that it had defined those two issues, both in detail in the earlier pages of the report and in the seven findings of fact. But, as with the commonality of Grant Newby's experience aboard Flights 301 and 304, deciphering and parsing a probable cause statement is speculation. The investigation being conducted carefully, thoroughly and within the parameters of all existing laws and regulations, the agency fulfilled its mandate in issuing a finding of probable cause. The case of Eastern Flight 304 was closed.

Although Grant Newby's Eastern Air Lines first officer's uniform jacket was recovered floating on the surface of Lake Pontchartrain the morning after the crash, Newby's body was not recovered and identified until March 18, some 22 days later. Along with Newby, positive identification was made that day of Captain William Zeng, Marie-Hélène Lefaucheux, and flight attendant Barbara Eubanks, and 11 of the passengers, including three-year-old Gabrielle Brisson, whose parents and brother were also onboard.[43] The other passengers and crew members were either fragmented, unidentifiable, or not recovered from Lake Pontchartrain.

Eastern Air Lines disintegrated in 1991 after years of mismanagement, plundering, strikes, a radically altered airline landscape, and a war in Kuwait.[44] A memorial cenotaph containing at least some of the names of those aboard was placed in the Garden of the Last Supper section of the Garden of Memories Cemetery in Metairie.[45] The spot is about five miles down Airline Drive from New Orleans International Airport's runway one, site of Eastern 304's final departure.

Eastern Air Lines Flight 304

Summary
Eastern Air Lines, Inc.
Douglas DC-8-21, N8607
Lake Pontchartrain, near New
  Orleans, Louisiana
Tuesday, February 25, 1964
Fatalities: 51 passengers, 7 crew; no survivors.

Report
Civil Aeronautics Board
Aircraft Accident Report
SA. 379, File No. 1-0006
Adopted June 27, 1966; released
  July 1, 1966

Fatalities

Crew
Captain William B. Zeng, Ringoes, NJ
First Officer Grant Robert Newby, 39, New York City, NY
Engineer Harry Idol, 39, South Farmingdale, NY
Purser Grover Wesley "Dusty" Flowers, 36, Hampton, GA
Stewardess Tove Emma Jensen, 24, Hapeville, GA
Stewardess Barbara Delane Norman, 21, Atlanta, GA
Stewardess Mary Ann Thomas, 21, Hapeville, GA

Passengers
Ackerman, Renate, Dusseldorf, Germany
Adams, Samuel Trustin, 66, Harahan, LA

Bowler, Jack Kerry, APO 23, New York City
Brisson, Gabrielle M., 3, Bainbridge, PA
Brisson, Marie M., Bainbridge, PA
Brisson, Peter V., Bainbridge, PA
Brisson, Timoteo V., Bainbridge, PA
Byczynski, Carl A., Detroit
Byczynski, Joan Marie, 20, Detroit (Eastern stewardess)
Byczynski, Sophie T., Detroit
Castle, Haskell Louiese Sherman, Falls Church, VA
Castle, Walter Francis, Falls Church, VA
Chown, Harriet Leehahn, Portland
Chown, Roger Murray, 54, Portland
Collins, Vernon Arthur
Cook, Beulah, Dalton, GA
Cook, Joe James, Dalton, GA
Cook, Noel, Dalton, GA

Donovan, Sgt. Sherman J., Jr.
Earl, Alan Lloyd, New Orleans
Gminski, Anne M., Wauwatosa, WI
Gminski, Louise A., Charlotte, NC (Eastern employee)
Gminski, Stephen M., Wauwatosa, WI
Jergenson, Larry Dean, 21, Chicago
Johnson, Frederick H., Falls Church, VA
Johnson, Mabert H., Falls Church, VA
Kane, Patrick Kelly, New Orleans
Kirschenheuter, Fred, 38, Chalmette, LA
Lefaucheux, Marie-Hélène, Paris, France
Meltzer, Louis G., Washington, D.C.
Meltzer, Natalie S., Washington, D.C.
Morgan, Jean M., Atlanta
Morgan, Joseph Warner, 48, Atlanta (naval officer)
Nassif, Clarice Mary, Montreal
Pawlowski, Carolyn B., Alexandria, VA (former stewardess for National Airlines)
Pawlowski, Harry J., Jr., Alexandria (Eastern Air Lines first officer)
Plautz, Ethel Adams
Plautz, Louis
Robert, Jules Charles, New Orleans
Rygaard, Ole F., Allentown, PA
Scharry (or Sharry), Sgt. E., Rose Pine, LA
Sereikas, Joseph G., Chicago
Sibley, Robert Hugh, 20, Yuma, AZ
Smith, Lynda Gail, 12, Ramsey, NJ
Smith, Sylvia Lee, Ramsey, NJ
Spaulding, Joan Grace
Spencer, Kenneth Lee, Los Angeles
Thorp, Virgil S., Charlotte (Eastern employee)
Truly, Harry Nicholas
Willis, Georgeanna J., Griffin, GA
Yoder, Richard C., Villa Park, IL

Part II. Falls City

11

Dusting Off a Dusty Little Airline: Braniff International Airways

But Mr. Braniff, those guys out there are on strike, sir. They're trying to shut us down!
  I understand that, Buford, but they're still my boys, and I don't want them to get sick. Look at them out there! You go get those things [cups of hot coffee] and go treat them like human beings out there and be nice to them.
    —1953 conversation between a Braniff office manager and
    founder and CEO Tom Braniff during a mechanics' strike

Usually, the story of a defunct American "trunk" airline is not all that compelling, except perhaps to ex-employees, business types and hobbyists. There is perhaps one exception: Braniff International Airways. "Trunk airline" was the generic name for U.S. common carriers providing scheduled air service to domestic and international destinations; it was first defined in the 1938 Civil Aeronautics Act. There were 16 airlines at the time that fit the definition: American Airlines, Braniff Airways, Chicago and Southern Airlines, Continental Airlines, Delta Air Corporation, Northwest Airlines, Pennsylvania-Central Airlines, Transcontinental and Western Air, Eastern Airlines, Inland Airlines, Mid-Continent Airlines, National Airlines, Northeast Airlines, United Airlines, Western Air Express, and Wilmington-Catalina Airlines.[1]

Since the 1938 act, the trunk carriers had changed names, merged, or collapsed. As of this writing, Chicago and Southern, Northwest, Inland, Northeast and Western Air Express ended up (after various convoluted meanderings between 1953 and 2007) within Delta; Transcontinental and Western (later TWA) was absorbed into American in 2001; Pennsylvania-Central (later Capital) and Continental disappeared (in 1961 and 2010 respectively) into United.

Braniff (1982), Eastern (1991) and Wilmington-Catalina (1941) all imploded; National was merged into Pan American, which itself disappeared in 1991. After nearly 100 years of drama in the air, the 16 originals are now down to just three: Delta, American and United.[2]

Given all this convoluted corporate history, why then is Braniff, out of the 16, the story which remains the compelling one revisited by books and magazines and business schools in universities around the country? Primarily because of two events: Braniff's reimaging of itself in 1965 continues to be both an extremely interesting marketing/advertising story, as well as being the perfect representation of the 1960s in commercial aviation—a time when the industry went from 1950s staid, conservative, red-white-blue to sexy "Coffee, Tea or Me?" It became exciting, Golden Age, exotic, and internationally colorful. And Braniff was also the first major carrier after the 1978 Airline Deregulation Act to strive for the highest success only to go bankrupt and cease to exist, providing much fodder for case studies and theses in business schools and executive boardrooms (and not just within the airline industry).[3]

Both the story of what Braniff called "The End of the Plain Plane" and its dramatic and similarly colorful 1982 implosion are recounted more fully elsewhere, particularly in John J. Nance's *Splash of Colors: The Self-Destruction of Braniff International*. The very successful 1965 marketing effort does give a significant glimpse into the state of the airline in 1966 when Flight 250 went down and an overview of the sea change and its impact on the airline's operations is worth a short recounting.

Braniff's roots are traced back to 1920s Oklahoma, a new state rich in petroleum wealth. Prior to its 1965 transformation, Braniff's history is fairly typical among businesses of the American Midwest. On June 20, 1928, in Oklahoma City, brothers Paul and Thomas Braniff created Paul R. Braniff, Inc., to provide air service between the state capital and Tulsa, 98 miles away. The name of the company came from the fact that it was Paul Braniff who had been enamored with flying for a number of years. In fact, his flight instructor was actually Orville Wright himself, who also issued Paul's pilot's license. Tom, an insurance salesman, helped his brother get the venture underway, lending his business expertise, and the first flight was operated on June 28 using a single-engine Stinson Detroiter carrying five passengers.[4]

By the end of 1928, some 3,000 passengers had been carried between the two cities. In hopes of additional success, the brothers added Dallas, Texas, as a destination (served via Wichita Falls, Texas), on February 11, 1929. But the company was losing money, and it was sold to St. Louis, Missouri-based Universal Aviation Corporation, a venture of Louis H. Piper, who designed and built what are now near-iconic private, single-engine aircraft. A merger with Aviation Corporation of Delaware (AVCO) and absorption of several small companies like Paul R. Braniff, Inc., resulted in the 1930 renaming of the combined company as American Airways, based out of Fort Worth, Texas. American Airways became American Airlines in 1934.

Even though the first Braniff air venture was a money-losing proposition, its sale to AVCO was quite lucrative for the Braniff brothers and gave them the money for a second try at success. Braniff Airways incorporated on November 3, 1930, and served the same Oklahoma City to Tulsa route as the first company, but a faster plane was used, the five- or six-passenger Lockheed Vega. Over the next two years, the airline added Kansas City, St. Louis and Chicago service.[5]

Crucial to the growth and development of the country's air network in the interwar period (1918 to 1941) were U.S. Post Office contracts. Passengers were fairly scarce, so carrying mail between various cities provided the difference between profitability and extinction for the airlines. Braniff did not have a postal contract initially. The brothers teamed up with several other small carriers and formed the Independent Scheduled Air Transport Operators' Association. The organization then did what all American corporations do: spent cash to lobby Congress to advance their particular business interest. Paul Braniff was one of many who testified at congressional hearings, urging that postal contracts be reopened for more competitive bidding. In 1934, the group's wishes were fulfilled and all existing contracts were cancelled and offered for rebidding.

Braniff's efforts paid off when the Postal Service gave the airline Revised Route Number 9 to carry mail between Chicago and Dallas on a route with stops in Kansas City, Wichita, Ponca City (Oklahoma) and Oklahoma City. In 1935, Braniff also secured Route 15 to carry mail within Texas between Amarillo in the north and Brownsville in the south with stops in Fort Worth, Houston, San Antonio and a few other cities.[6]

From that point, with the airline on a more solid business footing (based on the profitability of air mail), slow, steady expansion and progress commenced. Equipment-wise, Lockheed's twin-engine L-10 Electra was added in 1935 and Douglas DC-3s were added in 1937. DC-4s were used in the postwar period, followed by the DC-6 and DC-7 in the 1950s. With the dawn of the Jet Age, Braniff ordered Lockheed L-188 Electra prop-jets and Boeing 707–200 turbo-jets.

All of this heavy iron was used to service a continually expanding network. By World War II, Braniff had expanded far beyond its Oklahoma City-to-Tulsa roots, flying from Brownsville, Texas, in the South, up to Denver and Chicago in the North, and applications were pending for ambitious plans to serve the country coast-to-coast, from Los Angeles all the way to Boston (neither of which would happen until much later). The war saw the airline training pilots for the Army Air Force and using its fleet to supply services all the way to the Panama Canal Zone. Braniff's headquarters were moved from Oklahoma City to Dallas Love Field in the spring of 1942. The airline became "international" when a subsidiary company was formed in 1944 to operate flights within Mexico.

Postwar route authority was granted to Braniff to start service to Havana, Cuba, and several South American cities. With that came the difficulties of building Braniff's own navigation and communications systems and dealing with the political intrigues related to serving several South American countries. It was at this point that Braniff officially added the "International" to its name. Its all–Douglas fleet now stretched between Chicago and Buenos Aires and was branded "Conquistador Service Linking the Americas." The planes were liveried in red, white and blue and the company concentrated on providing competent, on-time service across its network. By the late 1950s, the route network was extensive, the airline was professionally managed (Paul Braniff left to "pursue other interests" in 1935; Thomas Braniff was killed in the crash of a Grumman Mallard as he was returning from a duck-hunting trip in 1954, near Shreveport, and was succeeded as the head of the company by Charles F. (Chuck) Beard, a long-time Braniff employee.

The Beard era at Braniff (1954–1965) was centered around the goals of growth of the network, keeping up with technological advances, particularly in aircraft, on-time dependability, and sustaining profitability. For the most part, the airline met with success. And that success, slow and steady as it was, made Braniff a target for corporate takeover by the mid–1960s.[7]

Dallas-based insurance holding company Greatamerica Corporation was looking to expand its portfolio by 1964. In an internal study, Greatamerica Chief Financial Officer C. Edward Acker identified both Braniff and National Car Rental as being among the worst of so-called "poorly managed" companies and ripe for takeover. The designation was probably in the eye of the beholder; Braniff seems to have put up decent numbers during the Beard era, from roughly 1953 to 1964. A typical measurement of airline financial health, revenue passenger miles, increased at Braniff from 550 million to 1.5 billion over that period, and the airline's one-cent-per-share earnings increased to $2.03. Over several of these years, Braniff was number one in on-time performance and became the 11th largest airline in the world.

Regardless, Greatamerica CEO Troy Post used Acker's assessment to justify buying 57.5 percent of Braniff's stock in August of 1964. He rewarded Acker's role in the takeover naming him executive vice president and CFO of Braniff. In 1965, Post brought in Harding L. Lawrence, executive vice president of Continental Airlines, to replace the ousted Chuck Beard. Braniff was about to experience over 15 years of excitement.[8]

11. Dusting Off a Dusty Little Airline

At exactly 8:30 a.m. on a late April morning in 1965, Harding Lawrence convened his first senior staff meeting in the Braniff boardroom in Dallas. John Nance recounts the speech Lawrence gave to his executives (and it became industry legend almost immediately as well): "There are no limits to what we can achieve, no limits to the greatness we can instill in this airline, but it depends on a dedication to hard work, long hours, and a complete rebirth of the company," he said. It was time to chart a new course for Braniff, the dusty little airline. But there was more. Lawrence "wasn't asking their cooperation, he was insisting on it." The team effort required intense dedication; anyone who wasn't up to it needed to "step aside and let those who want to join this program do so."

It was my-way-or-the-highway time. Either work like the boss (a notorious workaholic who would eventually have a penthouse apartment built above his office so he never had to leave), or "have a nice life!" Some chose to leave; others decided to stick out the ride to see where it would take them. There was opposition to having the old guard of Charles Beard swept away for something new and trendy and someone who had no connection to Braniff or its history or family. But Braniff would never be the same. Those who managed to stay on after the Greatamerica takeover would see dynamic, attention-getting, artistic and airplane-filling marketing campaigns, designed to not only give the dusty little airline a dusting, but also a stripping and redesigning and expanding that was unprecedented for the era.[9]

Having helped institute the "The Proud Bird with the Golden Tail" campaign and new livery at Continental, Lawrence this time wanted to go ever further. He set up a meeting with Jack Tinker Associates in New York to begin the process for rendering even the "Golden Tail" dowdy, dull and boring. The redesign effort would create a big bang and get people's attention. Whether it would get them into revenue-generating seats was another matter.

Lawrence was looking far beyond flying pretty aircraft, with pretty stewardesses, from Shreveport to Fort Smith. He was looking to be the world's largest airline, with a cutting edge fleet and an air of excitement that would generate the revenue needed for worldwide operations, as well as a fleet of the aircraft of the future, supersonic transports (SSTs). His eye was on a future where Braniff was the world's largest airline and offered round-the-globe flights on supersonic Boeing 2707s or Lockheed L-2000s or even BAC-Aerospatiale Concordes—Dallas to Paris in five hours.[10]

In the company's 1965 annual report, Lawrence's policies and vision for the immediate future as well as its long-term planning were set out. There were six areas of focus for the expansion of Braniff International, and some bright spots that indicated the boss's plans were already bearing fruit—first steps towards that gleaming SST-based future.

The report glowed: "Your management has been actively pursuing the acquisition of additional routes and following is a report on those under consideration:

1. United States–Caribbean–South America Investigation
Braniff is seeking improved access with non-stop privileges from major gateways in the U.S. to points served in South America. Hearings before the examiner were concluded in August. In October, the Civil Aeronautics Board's Bureau of Operating Rights recommended termination of existing interchanges and that Braniff be the exclusive U.S. flag carrier on the West Coast of South America. The examiner's initial decision is not expected until late in 1966.

2. Pacific Northwest–Southwest Service Investigation
Braniff is asking for authority to provide service between points in the Midwest, South and Southwest and the cities of Seattle/Tacoma, Portland and Salt Lake City.... The Board's Bureau of Operating Rights recommended extension of Braniff's route north of Denver to Seattle/Tacoma, Portland via Boise. The examiner's initial decision is expected in late Spring or early Summer.

3. Reopened Southern Transcontinental Service Case
In mid-year the Board again denied Braniff's application for a Dallas-Miami route.... A Board decision is pending.

4. Chicago-Toronto Route
The U.S./Canadian Bilateral Agreement was signed on January 17 this year, authorizing, among other things, U.S. flag routes between Chicago and Toronto. Braniff will seek authorization for the Chicago–Toronto route, on which a hearing is expected to be scheduled early this year.

5. Reopened New York–Florida Renewal Case
Braniff is seeking route authority between Miami and Tampa on the one hand and the northeastern cities of Washington/Baltimore, Philadelphia, New York/Newark, and Boston on the other. Hearings are expected to be held early in 1966.

6. Northeast-Bahamas Service Case
Braniff has filed an application seeking authority to provide service between its U.S. cities and the four Caribbean areas of the Bahamas, Puerto Rico, Jamaica, and the Virgin Islands.[11]

Harding Lawrence would be particularly known for the way he aggressively went after routes which became available for bidding after the 1978 law ending regulation of airline rates and routes. The airline had stacks of applications ready to go the first morning filing was accepted. The airline wanted 626 routes, more than any other airline. They won 67 of those requests. By Christmas of 1978, they had 16 new cities and 32 new routes to serve.[12] But that was later. As you can see from the 1965 annual report, Lawrence was always aggressive in his belief in expansion and would spend his years at the helm aggressively attempting to prod the CAB to action in Braniff's favor.

At the point Lawrence took over, there was a not unreasonable expectation that flying supersonic transports (SSTs) was just a matter of a few years away—and not just a fantasy in Harding Lawrence's head. It was the most logical step forward in airline evolution; almost every decade, airliners got bigger, faster, sleeker, more glamorous, and safer. The upcoming wide-bodies, such as Boeing's 747, were thought to be another step (and a short-lived one at that) in this inevitable evolution.

Even as late as 1969, Hollywood and others still believed in a Mach 2+ future. In Stanley Kubrick's *2001: A Space Odyssey*, a ground-breaking film in its own right, after SSTs would come near-space shuttles between earth and stations in earth orbit. And those shuttles (as envisioned by Kubrick and his crew) wore Pan American World Airways' blue globe livery and were served by Pan Am stewardesses in antigravity boots.[13]

Back on the real planet, glamorous Pan Am, with romantic destinations all over the world, was sure of both its lofty position and its future. Robert Gandt, in *Skygods: The Fall of Pan Am*, recounts how newly hired first officers were told on their first day at "the world's most experienced airline" in very matter-of-fact tones: "Congratulations, gentlemen. You are going to be SST pilots." First Mach 2, then the Moon.[14]

On June 5, 1963, President John F. Kennedy announced a competition to build America's first SST. Boeing and Lockheed entered the contest, eventually settling on their respective 2707 and L-2000 designs. Kennedy's announcement came after some prodding; Pan Am's enigmatic and iconic leader Juan Tripp, announced in May that the airline was prepared to buy Concordes from the Europeans if the Americans couldn't get their act together. Within a month, the competition was underway, subsidized with federal money. Boeing won the competition on December 31, 1966, but never got beyond building a life-size plywood prototype, almost a Potemkin SST, because by 1971 the whole concept of an American supersonic transport was dead, the victim of market forces, financial problems, soaring fuel costs, and environmental concerns. America's airline future would not be supersonic (except for the very wealthy and the very famous, along with a sprinkling of contest winners).[15] As of this writing, Boeing's mockup is displayed at the Hiller Aviation Museum in San Carlos, near San Francisco.[16] (Seattle's National Basketball Association team formed in 1967 during the height of the excitement over Boeing's SST. The team was named the Seattle SuperSonics or just "Sonics." It would win just one national championship in the next 41 years, before being sold and moved to Oklahoma City and renamed the "Thunder.")[17]

By the late seventies, just about the only one who was still holding onto the idea of an SST fleet in American airspace was Harding Lawrence. It was a dream he had been cherishing since his days at Continental. He wanted badly to make it work. So, in spite of the reality of SST economics and restrictions on flights which developed in the 1970s, Lawrence still believed. In 1979, he was finally able to realize the dream. An agreement with British Airways and Air France allowed for a Braniff Concorde service from Dallas to London or Paris via Washington Dulles beginning January 12, 1979.

But the service was almost as big an illusion as the Boeing 2707 mockup. Due to U.S. law, the FAA would not allow non–U.S. aircraft a certificate of airworthiness. A Concorde had to be "sold" to Braniff every time the SST was operated in the U.S. Exterior registrations would be covered up with tape to reflect the change of country. For instance, if the Concorde being used that day was registered F-WTSA (signifying French registration), white aircraft tape would cover up the F in the external signage and replace it with N, the letter signifying U.S. registration. Since it was now a U.S. aircraft belonging to Braniff, it could be given a certificate of airworthiness and operate the scheduled flights within American airspace. U.S.-approved registration paperwork was placed in the cockpit, while the British/French documents would be stored—in the front lavatory.

Other weirdness reigned. While Braniff crews (14 of their pilots were eventually fully certified for Concorde operations) flew the overland, subsonic segments between Washington Dulles and Dallas, insurers of British Airways' Concordes insisted that a British Airways captain and flight engineer had to be onboard as observers during the Dulles-Dallas and back trip. The service lasted for almost a year and a half and was terminated in May 1980 due to disappointing ticket sales. The Braniff subsonic segments averaged only a 20 percent load factor, far below the 40 percent Harding Lawrence declared was needed for break-even success.[18]

Concorde dreams were not the only thing coming to an end; the financial situation was worsening, and the airline was having difficulties absorbing the costs of its quick expansion during deregulation of routes and fares in 1978. At the insistence of Braniff's creditors, Lawrence resigned on December 30, 1980. In May 1982 (a year and a half after Lawrence's exit), Braniff made a quick trip to a bankruptcy filing and the airline that prided itself on getting its passengers there with "Flying Colors" faded into history. This was all in the future, however, in 1965. Lawrence had some dusting off and spiffing up to do first, not to mention all the expansions. No one would ever operate a B2707 between Fort Smith and Tulsa. But they could from Los Angeles to Singapore and Miami to Rio de Janeiro. It was just a matter of procuring the international glamour routes. Time to get to work.

12

Airstrip: "The End of the Plain Plane"

So this stewardess enters the cockpit and asks the captain, "Coffee, tea or me?"
He displays his best leer and answers, "Whichever is easier to make."
—Donald Bain, preface to the 2003 re-issue
edition of 1967's *Coffee, Tea or Me?*

"How was your flight?"
"Well, aeronautically it was a great success. Socially it left quite a bit to be desired."
—Noel Coward

The dust began to come off the little airline almost from the first moments of Harding Lawrence's first day at Braniff Airways. It would still be stirred up over 15 years later when he departed for good. But first things first. As Lawrence and his executives worked to get expanded routes and secure newer, better aircraft to serve them, he undertook one of the quickest and easiest ways to get attention and make a statement—a complete, modern, unusual and unprecedented redesign of everything from aircraft and uniforms to sugar packets and airport signage.

During the redesign of Continental Airlines, which resulted in the "Proud Bird with the Golden Tail" campaign with its bold new graphic and red-and-gold striping, Lawrence worked with Jack Tinker and Associates of New York. Tinker was also to begin preparations for promoting Continental's own planned SSTs. But when Lawrence decamped for Braniff and doubts about SSTs were mounting, Tinker agreed to accompany Lawrence to Braniff. The new CEO rewarded the agency with immediately lucrative work: shake up the airline, shake the industry and get people's attention. Make Braniff colorful, exciting, and hip. And do it yesterday. Tinker and his top executives, Mary Wells and Stewart Greene, immediately took up the challenge. (It was not the first time Wells and Lawrence had worked together professionally. Soon, they were also working together romantically and were married in 1967; their marriage would survive Braniff and countless other accounts, ending only with Lawrence's death in 2002.)[1]

But first there was the "End of the Plain Plane." Following their mandate, Tinker, Wells, and Greene began exploring color in every way they could imagine. In her 2002 autobiography, Mary Wells Lawrence said she saw Braniff "in a wash of beautiful color." She collected magazine pictures in the hundreds of colorful autos, interiors and clothes, while Tinker and Greene worked on coming up with ways to make Braniff's fleet come alive.[2] Mary managed a meeting with Italian fashion designer Emilio Pucci. He would produce not just colorful outfits, but also jaw-dropping, innovative (and occasionally impractical and downright sexist) outfits. But there was no question that the days of the dowdy stewardess were about to end. In fact, Braniff tossed out the title "stewardess" along with the old dowdy uniforms. Braniff's cabin crews were now "hostesses" entertaining colorful social soirees in sophisticated style at 36,000 feet.[3]

At the time, New Mexico artist Alexander Girard designed a restaurant named La Fonda del Sol (the Inn of the Sun) in the Time-Life building in New York City. The colors he used were brilliant, vibrant and suggestive of Mexico and South America, which would be increasingly a big part of Braniff's international destination expansion. Girard worked with designers from the Herman Miller firm for furniture and fabrics for La Fonda del Sol. Greene and Wells wanted his participation in the Braniff makeover, and they flew to New Mexico and signed Girard (who had known Pucci during Girard's years in school in Florence) as a project designer. According to Mary Wells, between the three of them, there were thousands of ideas for terminals, clubs, and luggage tags, and they were all eager to get moving.[4]

Jack Tinker was still working on the aircraft, the big, important piece. Various efforts came down to one simplified design by Greene, in just one color: green. Wells didn't like green, she claimed, so she asked Greene to make the same design in different colors. They first, she said, tried planes of all one color or the other. They then did one with all seven colors they were working with; it was "a thunderbolt" according to Wells; they had their "big idea," which would be "big news" bringing people from all over to get a glimpse of them.

More tweaking of colors and testing (on pieces of aircraft metal and a company DC-6) followed. At the same time, Girard worked on textiles, ceramics, signage, and seat covers. Pucci let his imagination run wild with uniforms for everyone from pilots to baggage workers.[5] All of this work was unveiled in 1965. Braniff ran a major ad campaign in magazines and newspapers of the day. One of the first new full-page ads highlighted all the color and excitement the redesign team was shooting for; ad directors put Braniff employees on the wings of a freshly painted Boeing 707, along with newly re-covered seats, indigenous people in native costume from countries served by Braniff, and all the accoutrements—pillows, blankets, silver service and everything else. The setup was used for a TV commercial as well as for the print campaign. The television ad ends with, "Braniff International. We won't get you there any faster ... it'll just seem that way."[6]

The print ad copy is instructive as to the redesign team's philosophy and excitement (not to mention that of Harding Lawrence himself) for what was being unveiled at airports nationwide. The copy is matter-of-fact and sophisticated:

> The first airplane was launched in 1903, but it stayed up for only 12 seconds.
> The next 63 years were spent perfecting it.
> Testing radar, mapping flight plans, creating a reliable and powerful engine—the jet—the airlines had neither time nor inclination for much else.
> An airplane had to fly. It didn't have to be beautiful.
> Planes were plain.
> And the regular passenger, the poor guy who spends much of his working time in the air, had to put up with this monotonous sameness day in and day out.
> For him, even at 600 miles per hour, flying was a crawling bore.
> The time had come, we decided, to add the last finishing touches to the airplane, and make it the nicest—not just the fastest—way to travel.
> We looked around for people who could take the idea and make it work. And we settled on two men.
> Alexander Girard.
> Emilio Pucci.
> Girard had designed a restaurant in New York, La Fonda Del Sol, and had won acclaim from critics and patrons alike. (As rumor has it, so many customers were walking off with the salt and pepper shakers that the restaurant had to replace them with less tempting models.)
> Pucci is something of a legend in his own time. An ex–Italian Air Force pilot turned fashion designer, he possesses an uncanny knack for making men look more like men, and women look more like women.
> Pucci redesigned our hostess uniforms, our pilot uniforms, even the uniforms of our ground crews.
> (When he unveiled our airline collection along with his regular designs in Florence, the airline clothing stole the show.)
> At the same time, Girard was busy redesigning our airplanes.
> Where airplanes had always looked like huge aluminum cigars with stripes down their sides, Girard selected 7 colors and painted the entire fuselage.
> (You can fly with our airline 7 times and never fly the same color airplane twice.)

> Where airplane seats were always covered with tasteless upholstery, Girard covered the interiors of our planes with Herman Miller fabrics.
> Again, 7 different interior designs. Seven different color schemes.
> Nothing was left untouched.
> Tickets and ticket offices were redesigned.
> Dishes and flatware.
> In-flight stationery.
> Our passenger lounges.
> The packages that hold the sugar for the coffee.
> Even the tissues in the lavatory.
> In little more than six months, Girard and Pucci initiated 17,543 changes.
> We have the most beautiful airline in the world.
> Braniff International
> United States | Mexico | South America.[7]

The response was positive, and the campaign would go on to a place of immortality with slogans such as "Plop plop fizz fizz, oh what a relief it is!" and "I Love New York!"—both from the agencies run by Tinker or Wells or both. The campaign is still beloved by many and is studied in higher ed advertising classes.

"End of the Plain Plane" got the attention of the press. *Time* magazine wrote a short article on it as well as Harding Lawrence's ascendance to Braniff's top spot. In their April 15, 1966, edition, *Time* wrote, "Colors are Fun," and listed all of the colors in which the planes had been repainted and the interiors redesigned and the hostesses redressed. The hostesses' uniforms were to be the most attention-getting of all the attention-getting redesigns at Braniff. Jokes were made calling hostesses "Pucci Galores." (The James Bond film *Goldfinger*, starring Sean Connery with Honor Blackman as a Bond Girl named Pussy Galore, was released in December 1964/January 1965. The joke basically wrote itself.) Bond girls were symbols of the sophistication and sleek, sexy style of the James Bond franchise. "Pucci Galore" as a tag for Braniff's hostesses wasn't that far (if at all) from the image the airline wanted to promote. Previous airline stewardesses were dressed in severe military styles; many of the earliest ones had to be nurses. But by the 1960s, the image was already slowly changing—Braniff and Pucci just gave it a large shove to kick it in high gear.[8]

As part of this renovation of the airline stewardess, Pucci's design was not confined to the cut or color of the uniforms. His concept was that Braniff's hostesses could have different looks for different situations. They could start all buttoned-up in coats, head-scarves and even plastic bubble helmets to keep out rain instead of tacky old umbrellas. But as she arrived at the airplane and greeted passengers, then proceeded through takeoff to meal service to drink service to arrival, she would progressively remove layers of the uniform until she was in her final configuration for landing.

Called the "Air Strip," this aspect of the redesign received plenty of press and attention, but sometimes the execution could be rather strange. The plastic bubble helmets to keep out wind and rain proved to be awkward to handle and store onboard. They could crack easily. And they would also be unlikely to be used for more than a minute or two, during a walk to or from the airplane to the terminal. As more and more terminals were using jetways with no need to be exposed to the elements, the plastic helmets became more and more useless. And with Braniff having a multitude of short routes in its system, the full Air Strip promised passengers didn't have time to be fully executed, leaving the hostesses at some weird stage of the process and the passengers reluctant to leave, wondering what they would miss next.[9] The basic idea was explained in print and television advertising when the uniforms were unveiled. On the print ad copy, the process is explained:

> Introducing the Air Strip.
> We had a girl go through the motions to show you just what's coming off at Braniff International.
> As to the picture below, our hostess appears at the airport wearing a reversible cold-weather coat, matching gloves and boots and, if it's raining, an ingenious plastic helmet.
> When she boards our airplane, she
> Zip
> sheds these outer garments to greet you in a raspberry suit and color-coordinated shoes.
> This ensemble is too expensive to risk soiling during dinner, so at the appropriate moment, she
> Zip
> Snap
> Zip
> changes into a lovely serving dress which we call a Puccino (named for its creator, Emilio Pucci, who believes that even an airline hostess should look like a girl).
> After dinner, our hostess
> Zip
> slips out of the Puccino, revealing the way-out outfit on the right.
> Each change is made in a flash, which allows her to give you constant attention, from the time you take off to the time you land.
> If the flight seems all too short, that's the whole idea.
> Braniff International
> Flies United States | Mexico | South America.[10]

Leaving aside the obvious sexism of both Braniff's Air Strip and the aforementioned Executives-Only flights at United, here was an airline more fully embracing the old concept in advertising: Sex sells. Jokes like "Pucci Galore" may have been meant by crew members of other airlines to make fun of Braniff, but instead, it just helped the sex to sell—something Braniff was not likely to admit publicly. Big sex appeal for the whole airline, not just one flight on one route per day with a few cigars a la United, sold big. In fact, for all the detractors (and they are legion) of both Harding Lawrence and Mary Wells over the "End of the Plain Plane" era, *Time* judged the pair, especially Lawrence, as having the "last laugh": "Braniff is getting more attention than other airlines, and operating statistics to show it. Passengers increased 18 percent last year to 3,370,000; revenues also rose 18 percent, to $129 million, and earnings were up 58 percent to $9,400,000. Within the year, Braniff stock rose from $25 to $125, and stockholders last week happily approved a two for one split."[11]

Sexy/sexist ads and pastel airplanes weren't all responsible for those financials. As *Time* noted, "Between 1945 and 1964, Braniff had slipped from fifth to ninth place among U.S. trunk airlines, was notorious for late operations, sloppy service and shoddy equipment. Its routes included everything from long marathons to Buenos Aires to costly Texas puddle jumps, but the airline had not won a new route for ten years and was barely making money."[12] It took more than $6.5 million for red, orange and green airplanes to fix all those shortcomings.

New aircraft came on order almost immediately to be available for future growth. And Lawrence did what Southwest Airlines has copied so successfully today; Lawrence had Braniff's Boeing 707s utilized for daytime flying and nighttime maintenance. Instead of just eight hours a day, the four-engined workhorse, largest in the fleet at that time, began flying 11 and a half hours, meaning more revenue and profits. The *Time* article ended with a recitation of Lawrence's fleet plans for the coming years. Noting orders for additional Boeings and a new British twin-engine liner, *Time,* like others in 1965, noted Lawrence's ongoing SST love, writing that he had "already put in an order for eventual U.S. supersonic transport planes, [and] may soon give a duplicate order for the British-French Concorde SST. Planemakers can hardly wait to hear what color Lawrence wants them painted."[13]

But the SSTs were still years away. In the meantime, how was Braniff to serve its existing money-making routes with better airplanes, preferably jets? Harding Lawrence went shopping.

13

Fastback Jet: The BAC One–11

> *Despite the belated arrival of a direct competitor, there is every indication that the BAC One-Eleven is well on the way to becoming the brightest star in the British aeronautical firmament.*
> —Aircraft Engineering: The Monthly Journal
> of Aeronautical Engineering, 1963

At the beginning of the 1960s, Braniff had the greatest percentage of domestic flights in short-haul routes (defined as being those with less than 250 miles, or 400 km, between city pairs) of any U.S. trunk carrier. As the jet age dawned, Braniff looked for a suitable aircraft to replace the aging DC-6s, DC-7Cs and Convair-Liners in the fleet with larger capacity, faster aircraft.

Braniff believed it found what it was looking for in British Aircraft Corporation's One–11 jet. On October 13, 1961, Braniff became the first United States airline to order the British product, inking an order for 14 of the 65-seat twinjets. Charles Beard and the old Braniff made the forward-looking choice, but the first Braniff One–11 would not be delivered until March 11, 1965, after Harding Lawrence took over the company and plans for the "End of the Plain Plane" were well underway.[1]

In the company's 1965 annual report, Braniff was still focused on its destiny: world-wide supersonic service. But it also acknowledged the airline it was as well as what it was becoming. The report to shareholders states:

Braniff now has a 10-year equipment plan that will take us well into the supersonic (two-thousand-mile-per-hour) era. As a result we have been able to accelerate our timetable for transition to subsonic jets.

In looking at equipment one must understand Braniff. It is really four airlines:
—A short-haul airline that connects to itself and other carriers in the Southwest;
—A regional airline that serves the expanding needs of the Midwest;

—A long-haul trunk carrier operating between New York, Washington, Chicago and Texas;
—An intercontinental carrier serving Mexico, Central and South America.
To solve our jet equipment needs, we have purchased equipment that will best serve each of our four airlines.[2]

Three of these four airlines were pieces of other airlines as well as the original Braniff, and the fourth was the foundation for Lawrence's grand futuristic vision.

Braniff, the original core, was the "short-haul airline" serving the Southwest. The regional line serving expanding needs of the Midwest was the remnants of Mid-Continent Airlines, merged into the old Braniff in 1952. And the intercontinental carrier serving Mexico and points south was Braniff's Latin American Division, which would soon be strengthened by merging into Braniff International, beginning in 1967. The last airline mentioned, the long-haul trunk carrier between New York, Washington, D.C., Chicago and Texas, was partly old Braniff and partly future Braniff: future routes would serve bigger destinations like Los Angeles and Boston and Seattle and Miami, as long as the CAB could be convinced to help make it all happen. By the 1970s, Braniff executives expected to have the four-airlines-within-an-airline better integrated and served by new jet equipment. And there was the old dream of Braniff SSTs running celebrities between Los Angeles and New York in three hours several times a day.[3]

So that the dream of order could be brought to reality out of all this chaos, it wasn't enough to spice up planes and hostesses. Braniff had to make wise choices in what aircraft and how many were needed for its system. It needed a plan and the money to make it happen. The report went on to note the choices that were made for each of the four components of Braniff. For the regional operation serving the midwest (the old Mid-Continent routes), the airline "ordered twelve Boeing 727 QC aircraft. The 'QC' stands for 'Quick Change.' Braniff will be the first airline to introduce these convertible tri-jet planes. This means simply that during peak passenger hours they will have a maximum passenger capacity, whereas during off-peak or night periods they can be converted in 30 minutes to full-cargo capacity or a combination passenger/cargo configuration."

For the long-haul trunk portion of Braniff, the airline would stand pat: "we will continue to operate our ten Boeing 707's and 720's, a completely competitive and comfortable fleet for this requirement." In addition, Braniff also "ordered five intercontinental Boeing 707–320 C's. This fan-jet fleet will be used to supplement our 707's on our international routes, giving us more capacity and improved performance. They will also be utilized to fulfill our Military Airlift Command (MATS) commitments commencing in June. In addition,

13. *Fastback Jet*

British Aircraft Corporation One-11 N1553 of Braniff International Airways, taxiing at Dallas Love Field in 1966. Painted in the ochre color of the "End of the Plain Plane" campaign, ship 53 disintegrated in turbulence over Falls City, Nebraska, August 6, 1966, while operating Braniff Flight 250 (Christian Volpati).

Photographs of Braniff's N1553 were fairly rare since it was only in the air for eight months, therefore almost brand new. It is seen here at Dallas Love Field in the summer of 1966 with one of Braniff's Boeing 707s at left, both sporting Alexander Girard's distinctive lettering (photograph by Bob Proctor/Jon Proctor Collection).

these planes also have complete convertibility. Cargo capacity can be increased or decreased according to passenger and cargo demand."

For the short-haul routes, Braniff was taking delivery of the 14 British Aircraft Corporation One–11 planes during 1965: "These rear-mounted, twin-jet aircraft transport 63 passengers quietly and comfortably between our small and major connecting cities. We were the first to have short-haul jets. These versatile planes have provided eight of the cities we serve with their very first jet service and have permitted increased service to many of our other cities." Finally, the airline announced that the end of the Piston Era was near: "The new jets, when added to our present ten 707's and 720's, will give us a fleet of 41 pure jets and the transition to turbine power will be completed. Ten DC-6 and three Convair aircraft have already been removed from service and sold. Our five DC-7C aircraft and more of the Convairs will be taken out of service this coming year and replaced by jets."

The annual report noted some flexibility in its 10-year plan: "Under current review within our 10-year plan is any future requirement for additional subsonic jets to meet forecast passenger and cargo traffic growth and any need for supersonic jets beyond the present two positions that Braniff has reserved for the American supersonic transport." Braniff in 1965 was still planning to fly some sort of SST as soon as possible.[4]

In order to afford the expansions and supersonics of the future, Braniff needed a sound financial footing in the present. Bread-and-butter routes to cities such as Omaha and Kansas City needed to be served with the best, most profitable aircraft available. Over the years, the industry, regulators and financial analysts developed several measures of airline financial health. The Airline Data Project of the International Center for Air Transportation at the Massachusetts Institute of Technology[5] put forward some useful definitions of these measurements. The "basic measure of airline passenger traffic" is revenue passenger miles (RPMs). This measurement shows "how many of an airline's available seats were actually sold. For example, if 200 passengers fly 500 miles on a flight, this generates 100,000 RPMs."[6] A second measurement is available seat miles (ASMs). One ASM equals "one aircraft seat flown one mile, whether occupied or not. An aircraft with 100 passenger seats, flown a distance of 100 miles, generates 10,000 available seat miles."[7]

Passenger load factors use the previous two measurements to "represent the proportion of airline output that is actually consumed." Load factors are "Revenue Passenger Miles (RPMs) expressed as a percentage of ASMs, either on a particular flight or for the entire system." Load factors for a flight or an airline are figured by dividing RPMs by ASMs. "Load factor for a single flight can also be calculated by dividing the number of passengers by the number of seats."[8] There are many other acronyms and terms by which an airline's financials are sliced and diced, but that goes beyond the scope of this narrative. Where Braniff and its aircraft are concerned, load factor is the focus.

In the 1960s, the average load factor for U.S. airlines (as noted in Department of Transportation data) averaged 53.88 percent—the airlines sold just over half of their available seats. As a comparison, load factors have steadily increased since fare and route deregulation in 1978. Load factors for all carriers in 1995 stood at 67 percent and rose to 83 percent by 2011. Available seat miles for all airlines operating domestically in the U.S. were 614,610 in 1995 and 698,076 in 2011. So while the number of seats rose fairly modestly, the number of those available seats sold was much higher during that time period. While there have been ups and downs, the airlines would seem to have figured out how to get more passengers into the seats they have to sell.[9]

As Harding Lawrence himself reportedly said (or it could have been Groucho Marx; the quote has been attributed to several people in and out of the industry), an airline seat is highly perishable; once the aircraft door is shut and the flight leaves the gate, an empty airline seat can never be resold. Putting a person in a seat is an imperative for an airline; one person in one seat on one flight can even be the difference between making a profit or sustaining a loss on a trip (which is perhaps an oversimplification).

While tastes have changed over the years (many people no longer care what haute couture the flight attendants wear and couldn't tell the difference between Alexander Girard and Alexander McQueen), the problem today continues to be what it was in the 1960s: selling the seat while holding down costs. Higher costs mean the airline has to sell more seats in order to break even or turn a profit. And that's where load factor comes in; an airline will know, for a specific flight with a specific airplane, how many seats it must sell to turn a profit. If it has a 100-seat jet on a 500-mile trip and it needs to sell 50 seats to break even, the airline will be very happy when seats 51–100 have passengers in them when the aircraft departs the gate.

For the ten-year period 1960–1969, Braniff's load factors of 53.03 percent were in line with the industry average of 53.88 percent. Braniff turned in similar numbers throughout the decade, from 56.32 percent in 1960 to 56.10 percent in 1966 (the first full year of Harding Lawrence's tenure) to a low of 46.06 percent in 1969.[10] Controlling costs is an important part of this complex balancing act. An airline like Braniff in the 1960s needed an aircraft which would allow it to be profitable at load factors of 53 percent. When an aircraft manufacturer tells a potential customer that its aircraft can reduce the load factor needed for profitability on certain routes, it gets the airline's attention. British Aerospace Corporation did exactly that when it offered its One–11 twinjet to Braniff.

BAC conducted internal cost calculations for the One–11 in 1960. Their data showed that the One-11 configured with 74 seats "would break even over 100 statute miles with 53 passengers (71.8 percent load factor); 40 passengers (54.2 percent) over 200 statute miles; 31 passengers (43 percent) over 400 statute miles; and 28 passengers (37.8 percent) over 800 statute miles." The numbers would change during development, manufacture and deployment, but in them Braniff found a compelling case for using the British jets on its short-haul routes.[11]

BAC was formed in 1960 from four different corporate aircraft designers/builders: English Electric Aviation, Vickers-Armstrong's aircraft division, Bristol Aeroplane Company and Hunting Aircraft. It was Hunting which began work on what would eventually become the BAC One-11 earlier in the 1950s. At first, it was conceived as a jet replacement for the popular Vickers Viscount turboprop, which was used all over the world.

Vickers sold 445 Viscounts, including to airlines in the United States. Continental and Northeast flew the Viscount, but it was mainly associated with Capital Airlines, which used it throughout the 1950s. Capital's financial position deteriorated in 1960 and Vickers threatened to foreclose on the fleet; Capital's merger into United Airlines saved the Viscounts from that fate. They flew on in United colors until 1969. The Viscount's jet replacement, as envisioned by Hunting, had 30 seats. Development proceeded in fits and starts until the 1960 integration into BAC. After the integration, designers were able to draw on Vickers' institutional experience with the Viscount to develop the new short-haul jet.

The mantra was for the new liner to do anything the Viscount could do but do it 200 miles per hour faster and at slightly lower operating cost. Faster flights equaled more flights, which, coupled with lower operating costs, meant more profit. The One-11 would be able to fly into the same airports as the Viscount because its landing speed was designed to be slightly slower.[12]

By 1963, the jet's configuration and design was set. It would be in varied seating configurations with room for up to 74 seats in a one-class layout. To help boost profits, the jet was "built to spend its life in the air," since every minute of time on the ground represented the unavailability of its seats for sale. It's not only true that an empty airline seat is lost forever once the flight leaves the gate, it's also true that airlines seats don't produce revenue while they're sitting empty on the ground. To boost fleet utilization time, the One-11 had built-in features designed to shorten time on the tarmac. It had crew-operated boarding stairs (no need to wait for loading bridges or portable stairs) and an auxiliary power unit which could operate air conditioners, engine starters and other equipment, without the need for ground crews to connect/disconnect portable generators. The aim was to cut the turnaround times of turboprops in half.[13]

BAC also touted the adaptability of the interior of the aircraft, which could be changed in a heartbeat "from a 79-passenger layout to an arrangement catering for the most discriminating private party."[14] At first blush, it might seem a bit far-fetched for an airline to earn much revenue hiring out its aircraft for private parties. But given Harding Lawrence's penchant for big bashes, it wasn't quite as far-fetched in Braniff's case as it might seem.

The big advantage was jet engines. In this case, Rolls-Royce supplied two Spey turbofans capable of driving the aircraft to 540 miles per hour. The engines were placed at the rear of the aircraft instead of on the wings. Sud Aviation's Caravelle twinjet was the first to use this type of arrangement; the One–11 would be the second. Douglas's DC-9 and Boeing's 727 would follow the trend, which is still alive today in a variety of airliners now known as "regional jets." In fact, the One–11 can be described as being one of the first regional jets, if not the first. It served short to medium stages and carried less than 100 passengers. The One–11 offered large airliner-type interior space, with up to five-abreast sitting in coach.[15]

British Aircraft emphasized the integrity of the One–11's technical and structural systems. The company's four factories used "integrally machined components for smooth load distribution and ruggedness with high standards of safety and minimum weight." The company aimed for long life and low airframe maintenance expenses. This "ruggedness" was particularly important in short-haul aircraft, which take-off and land (and pressurize/depressurize) frequently. As bitter experience with the de Havilland Comet series of aircraft had shown, repeated pressurization cycles could lead to structural failure in improperly designed/manufactured aircraft.[16]

The Comet was the first passenger jet airliner, which began flying with BOAC on May 2, 1952. In May 1953, one of the company's Comets crashed into the Indian Ocean after takeoff from Calcutta, India. In January 1954, another Comet was lost over the Mediterranean after takeoff from Rome. Courts of inquiry and a major test program revealed structural failures (largely from metal fatigue resulting from the pressurization cycle) in both aircraft, especially at the corners of the liner's square windows. The Comet was partly redesigned and strengthened with thicker gauge metal in the fuselage and wings. The square windows were replaced with round ones.[17]

As it was, the Comet's experience was tragic but invaluable. British aircraft manufacturers benefited from the knowledge gained, as did Boeing, Douglas, Lockheed and General Dynamics in the U.S. It was the Comet tragedies that led to BAC laying so much stress in its promotional materials on the strength of design, materials and construction in the One–11. This aspect of the One–11s history is important to the later narrative of Braniff Flight 250. British Aircraft believed in the integrity of its jetliner and took nervous first steps to determine what brought Flight 250 down. Even the hint that another Comet tragedy was on their hands made them understandably nervous.

BAC's marketing brochure reassured potential customers on this score: "The BAC One-Eleven makes considerable use of milled-from-the-solid wing planks and fuselage panels. The skin milling shop at Weybridge, the finest in Europe, is one of four such facilities in BAC which are employed on One–11 production. Same as on the Viscount and VC10."[18] BAC included a structural drawing in the brochure showing that the inner wing core structures and fuselage skin, plus the wing spar box, were constructed of light alloy metals; the engine mount system was of steel. The fuselage skin was wholly made from "single, heat-treated, copper-base aluminum alloy." This choice was made to "give both good life without cracking and [to] slow crack propagation should cracks begin to appear." BAC's testing showed the alloy "demonstrated long life and slow crack propagation rates even with large cracks in the skin and the whole of an internal frame cracked." This way, cracks could be located by external examination without concern for the internal frames. They used tests on the Vanguard to demonstrate that "crack propagation slowing" worked; crews in the air or on the ground could see that something was wrong without its resulting in a Comet-like catastrophic loss of the aircraft and its passengers.[19]

In a 1964 edition of *Aircraft Engineering Journal*, a highly detailed, 32-page section was devoted to One–11's design and manufacture, and its strength, safety and durability were repeatedly discussed. For instance, the article goes into great detail about the One–11's wings. Designers created five machined frames attaching the wing to the fuselage, it said. "Bending strength was maintained by a keel box running below the wing." The "junction where the outer wing is attached to the centre-section wing joint" was a "proven design" and was used on the Vanguard and VC10. Loads were transferred across this joint by "double butt straps and bolts in double shear, interference fits being used in tension areas. The butt straps are made in relatively small pieces so that cracks occurring in one cannot spread very far." All of this "engineering talk" shows that the wing's design was already proven in flight and that additional safeguards were built in, including components which, in failure, would not, like a row of falling dominoes, cause other components to fail as well.[20] There was some deviation from the previous learning, however: "Fuselage skins above the wing depart somewhat from past practice. This region is heavily loaded in shear and needs considerable stiffener support ... by machining the stringers and skin from the solid, in the same way as the wing, a considerable reduction is made in the number of parts and in the assembly time in this area."[21] The fuselage was also stronger.

The design and construction of these two areas, the wing and the fuselage, are highlighted at this point in our narrative because of what would happen to a One-11 over Falls City. Given the aircraft's design and manufacturing concept based on both good (Viscount/Vanguard/VC10) experience and bad (Comet), BAC's highlighting of the jet's great strength was not just a figment of the marketing department's imagination. Designers and builders (as well as airline customers and their pilot and maintenance teams) believed in that strength. As will be seen later, that belief was not necessarily misplaced but was perhaps hubristic. "Fail safe" implies "safe from failure," and is comforting, even when we know that everything (even the earth itself and stars in the sky) eventually fails—or at least is changed into something else. Engineers, particularly of the aeronautical variety, try to make products that will either not fail unless it's on purpose, or that will not injure or kill humans when failure occurs. The One-11 (and other planes like it) was supposed to be destructible only if humans destroyed it.

By 1964, British Aircraft, building on its history and experience, produced an impressive aircraft in the One-11. The company also furthered its technical knowledge, which would become invaluable when it began work on its joint project with France's formerly named Sud Aviation: the BAC-Aerospatiale Concorde supersonic transport.

It's worth mentioning that, after so much work, hype and hope on the part of small boys and Pan Am pilots, Braniff's CEO and Hollywood movie stars, the world's supersonic future finally began January 21, 1976, and it had the British Aircraft Corporation name on its fuselage. Unfortunately, by that time, Concorde was the only SST that had survived to see airline service. And only 14 of the type flew—for just two airlines, Air France and British Airways.

Concorde would score impressive performance and admirable safety records; the type operated for 24 years without a fatality until Air France Flight 4590 crashed near Paris in 2000 with the loss of 100 on the flight and four on the ground.[22] But even with a post-crash investigation, upgrading/strengthening and exhaustive testing of Concorde, her days were still numbered. After a final Mach 2 passenger flight on October 24, 2003, both airlines' SST fleets ended up in museums scattered around the world. Supersonic transports, once considered a permanent feature of the future, are now permanent exhibits of the past.

The first order for the One-11 came from British United Airways in 1961. Other orders followed over the next several years, including from Mohawk and American in the U.S. Eventually, 262 One-11s (58 if them Series 200s) would come off the assembly lines at Hurn and Weybridge (the last one on July 8, 1969). Fourteen of those were destined for that dusty little Texas airline, Braniff International Airways.[23] In the same article in *Aircraft*

Engineering Journal stressing the One-11s safety and durability, there is a small section devoted to Braniff's order. It was said to be remarkable for being the first time that a British airliner had been ordered by an American airline while it was "still at the drawing board stage."

BAC and Braniff said the historic order was based on "economic appeal," that the twin-jet was "specifically tailored to satisfy the requirements of the short-haul operator."[24] The economics of this deed having been already discussed above, it's helpful to take a closer look at the day-to-day reality of Braniff's short-haul routes. Braniff International Airways northbound Flight 250 and southbound Flight 251 were marathon runs up and down the country's spine during the mid–1960s, modified remnants of old routes flown by Mid-Continent before the merger.

Flight 250 originated at New Orleans' Moisant Field (the starting point in 1964 of the ill-fated Eastern Air Lines Flight 304 and now called Louis Armstrong New Orleans International Airport) at around 6:30 each evening and terminated at Wold Chamberlain Field in Minneapolis at 1:48 a.m., thanks to short stops in Shreveport, Louisiana; Fort Smith, Arkansas; Tulsa, Oklahoma; Kansas City, Missouri; and Omaha, Nebraska.

Its sister Flight 251 took the trip in the opposite direction; beginning in Minneapolis at 12:40 a.m., 251 flew south to New Orleans, arriving at 4:47 p.m. in the afternoon. Its stops were slightly modified from those of 250; the intermediate stations were Des Moines, Iowa, Kansas City, Tulsa and Shreveport, but it did not serve the northbound Flight 250's Omaha and Fort Smith stops.[25] These two flights were highly representative of the airline.

On October 13, 1961, Braniff became the first United States airline (and second overall) to order the British product, inking an order for six (which was eventually increased to 14) of the twin jets, which were configured for 63 passengers. There would be 24 first class seats up front and 39 tourist in the back.

After so many years of design, testing and manufacture, deliveries began in late 1964 and service was implemented as soon as each jet could be brought on-line. The twin-jet was officially designated the "BAC One-11" by Braniff, and the "BAC 1-11-203AE" by the manufacturer. Braniff's One-11s were given sequential American registration numbers N1541 through N1554. The first Braniff One-11 off the assembly line in England, N1541, had its first flight on September 6, 1964, and was delivered on October 8. (The last was delivered December 22, 1965.) Braniff quickly implemented the jet on the route served by 250 and other similar flights, such as Fort Smith to Tulsa (120 miles) or Kansas City to Omaha (185 miles). Cities such as Lubbock, Amarillo, Wichita Falls, Dallas, Shreveport and Colorado Springs would be among those getting One-11 service.[26]

The One-11s were given the full "End of the Plain Plane" treatment as soon as was practicable, but initially they carried the original red/white/blue exteriors. The interiors as delivered were designed by Charles Butler Associates of New York. An article in *Flight International* of April 22, 1965, described the original, pre–Harding Lawrence look of the passenger cabin: "Areas of seats are upholstered in blocks of coral, gold and turquoise blue.... On the cabin side wall there is a silk-screen pattern in a block print of muted tones of gold and various shades of grey, soft blue and coral which harmonize with the seats, carpets and curtaining. Called 'Carib,' this pattern is described as being 'borrowed from the influences of the Aztecs and Incas and tempered with the North American touch.'"

The top of the cabin and the overhead racks were "a beige golden tan" and the bulkhead consisted of two "decorative plastic laminates, one a cool ice-blue with silver and gold threads and the other a warm coral rose with a lace-like texture." Carpeting was of a predominant turquoise blue "flecked with all the cabin's colour tones." The aircraft's lavatories were "panelled with an unusual plastic material having the appearance of a soft yellow straw fabric."[27] As new, classic and restrained as the delivered interiors may have been, they were ripped out within months for the bolder "mod-ness" of the new regime.

A common practice of the onset of the golden era of commercial airline service was to designate aircraft with airline-specific marketing names, in spite of what the manufacturers called it. Thus, the Martin 4-0-4 twin prop liner became Eastern Air Lines' "Silver Falcon," and the company's DC-8s (such as that used for Flight 304) were called "Golden Falcon Jets." All of Pan Am's aircraft were called Clippers (and had individual names following, as in Clipper *America*). The Convair 880 became the "Starstream 880" when put into service by TWA. Delta had its "Royal Crown" DC-7s.

Eastern continued the tradition 20 years after the "Golden/Silver Falcons" by renaming the wide-bodied tri-jet Lockheed L-1011 Tri-Star the "Whisperliner." American Airlines designated most of its aircraft regardless of type as "Astrojets" during the 1960s and early 1970s; its DC-10s were called "Luxury Liners" as the decade went on. Even pre–"End of the Plain Plane" Braniff used the term "Jet-Power Electra" for its turbo-prop-powered Lockheed L-188 Electra IIs during the early years of the Jet Age, apparently trying to convince its passengers they were flying a "jet"—which just happened to have propellers.[28]

So it was in this tradition that Braniff's marketing department highlighted the BAC One-11's "clean-wing, rear engine design" by rechristening it the "Fastback Jet" in 1964.[29] That name, along with the original red-white-blue livery, would get swept out within a year as Harding Lawrence took the broom to his dusty little airline.

14

Happy Birthday! Bill Schock and That Night in August

*But, like the feller says, I wouldn't do it again for a billion bucks.
And, on the other hand, I wouldn't take a billion bucks for what
I've gone through. They just have to be the greatest experiences of my life.*
—Bill Schock, *Thrills, Chills and a Spill*, 2006

On a gray October morning in 2012, I drove from Omaha down to Falls City, Nebraska. The day's goal was to start at the local newspaper, the *Falls City Journal* and find out more about the events of August 6, 1966, the night Braniff Flight 250 fell to earth northeast of town.

After a short explanation at the front desk, I met the publisher, Scott Schock. Everyone in the office was friendly and helpful. Strangers wandered in from time to time asking about the Braniff crash. It's not too unusual for them. Scott listened to my explanations ("I'm writing this book...") and offered to let me look at anything they had that would help. He also asked if I wanted to meet his father, who was not only the paper's owner, but also the reporter on the scene the night of the crash. Absolutely! We walked into the office next to Scott's.

Sitting at a desk typing away on the next edition's Hanging out the Warsh column, was Scott's father, Bill, who also graciously stopped work to help me out. The first thing I saw above Bill Schock's desk was a black-and-white photo of a B-17 on a bombing run. Always a sucker for a B-17 picture, I trailed off in mid-sentence and stared. This one was a bit different. More than half of the tail was missing, but the thing was still flying. Bill followed my gaze, smiled, and said, "Isn't that something?" It turns out that a member of Bill's B-17 crew, Les van Gorkum, took the photo during a bombing raid over Berlin, Germany,

14. Happy Birthday!

in early 1944. The unlucky bomber, a B-17G-10-DL named *Silver Dollar*, was flown by Lt. Merlin Reed and his crew; eight of the men aboard died, and the other two were made prisoner.[1] It's an amazing picture and almost made me forget Braniff Flight 250.

But to the left of the B-17 picture, and lower on the wall, was another black-and-white photo. At first, I didn't pay it much attention, thinking it was another war picture, showing the devastation of wartime bombing. I asked him something about the Braniff crash. He pointed to the picture on the left and said, "Well, that's from the crash right there." This photo he took himself. I stood up and leaned in for a closer look. I'm not sure what I was expecting. In many crashes there is at least something recognizable; perhaps I was expecting to see a section of fuselage or something. But here was almost complete disintegration. Looking closer, I saw two hideously bent Herman Miller–upholstered seats (one striped, one solid) were upright among the debris. Other, similar, cushions were scattered around, some in stripes, some in checks. After always seeing Alexander Girard's designs in sophisticated 1960s color ads presented as the epitome of comfort and style, this was almost an obscenity.

Bill looked through a bottom drawer; he keeps copies of pictures in a folder to show people like me who randomly show up and ask. I flipped through them and it got quiet. Bill has seen this scene many times and he knew to wait. Most of them were either never published or printed only once in the *Journal* itself at the time of the crash. In fact, there are almost no pictures of the Braniff Flight 250 accident scene circulating. The NTSB report has a few drawings, as does Macarthur Job's excellent *Air Disaster* (Volume 1), which devotes a chapter to 250. The Falls City visit was the first time I saw the pics; that they are the originals is extraordinary. While "extraordinary" describes the pictures, it also applies to the photographer, even if he wouldn't like the description. Bill Schock's experiences as a prisoner of war in Germany from 1944–45 are recounted in his autobiography, *Thrills, Chills, and a Spill*. It's riveting reading and highly recommended.

But then came August 6, 1966. With the war far away in the rear-view mirror, Bill and the *Falls City Journal* were chugging along. It was a Saturday night and Bill and his friends could relax from the usual long work week. This particular moon-flooded night was special, however; it was Bill's 50th birthday, a milestone. So, friends and family went out to the Falls City Country Club on the west side of town to celebrate. Even though not much of a drinker, Bill "celebrated a little bit," and was therefore understandably a little confused when someone hurried up to him around around 11:30 p.m. a bit out of breath.

"Bill, there's been a plane crash out northeast of town!" Bill waved his hand a bit; it was late Saturday night, it was probably a little two-seater of

In the last summer before World War II, many Falls City, Nebraska, men joined Company B of the 134th Infantry. This group photo taken in 1941 at Camp Robinson, Arkansas, shows Sgt. Bill Schock (bottom row, third from the right). Schock left the infantry in 1942 to join the Army Air Corps and its mighty Eighth Air Force (National Archives).

some sort, and even with the full moon, they weren't likely to find any crash site until morning, what with the rolling hills and woodlands out towards the Missouri River.

"No, Bill, it's big. An airliner! You need to get out there!"

"A good newsman always has a camera in the car in case something happens, even if it's his own wedding or funeral. He and several others hopped into their cars and headed back east towards town.

"The word was that something had crashed out on Tony Schawang's farm. They arrived there while fires were still being put out. It was just about midnight. It was a surreal way to end a 50th birthday. Bill Schock pulled out his camera and started doing his job."

15

Milk Run: Braniff International Flight 250

> *In that obscurity the pilot's head and shoulders were all that showed themselves. His torso was a block of darkness, inclined a little to the left; his face was set toward the storm, bathed intermittently, no doubt, by flickering gleams. He could not see that face; all the feelings thronging there to meet the onset of the storm were hidden from his eyes; lips set with anger and resolve, a white face holding elemental colloquy with the leaping flashes ahead.*
> —Antoine de Saint-Exupéry, *Night Flight*

Moisant Field, Kenner, Louisiana
6:35 p.m. CDT, August 6, 1966
Saturday

The angry year of 1966 still had five months to run as BAC One–11 ship 53 baked on the tarmac at New Orleans Moisant Field. August 6, 1966, was a hot, steamy, typically New Orleans afternoon. The twinjet was being prepared for the northward run of Braniff International's Flight 250. Manufacturer's line number 070 and U.S. registration number N1553,[1] the One–11 was painted in the airline's new livery, in its case, a brownish-tan officially called ochre.[2] (There would be an unsubstantiated rumor years later that the color, taken from no-longer-used cans in a Braniff maintenance hangar, was used by Southwest as the basis of its livery, renaming it "Desert Gold." Whatever the provenance, the colors are indeed similar.)

Ship 53's first flight was, according to the NTSB, on December 8, 1965, with Braniff taking delivery shortly thereafter. The airframe and engines therefore had low usage time on them; the One–11 showed a total flying time since delivery of 2,307:38 hours and 2,922 total landings when it departed Kansas City on its last trip.

Two Rolls Royce Spey Mark 506/14/15 bypass turbojet engines gave N1553 a distinctive growling whine. The number one engine on the left side of the plane had flown a total of 3,122 hours, 1,984 since its last overhaul. The number two engine on the right side was much newer; it showed a total time of 237:38 and had not been overhauled. Obviously, the airline experienced the need to replace the original number two engine somewhere around 2,900 hours. Records do not indicate why this change was made.[3]

In fact, the airplane was still in near-pristine condition. Maintenance was requested at some point during the day before N1553 was due to operate Flight 250. A pilot on a previous flight that day reported that the "yaw damper jerks rudder." (Roughly speaking, the yaw damper helps smooth out the side-to-side and rolling motions which can develop in flight. It makes things smoother for the passengers.) The pilot made a statement to the NTSB investigation later that the problem occurred during the climbout to altitude after takeoff and was felt only through the rudder pedals. The discrepancy caused no directional problems and disengaging the primary damper caused the vibration to stop. Since this was not a crucial item for safe operations and there was a backup yaw damper in the autopilot system, the primary damper was not a mandatory fix and the aircraft could be dispatched without repairs and unnecessary delays that would spread up through the system all the way to Minneapolis. The NTSB, the FAA, and Braniff, as well as the pilots flying the aircraft, had no problems; Braniff mechanics would fix the problem when N1553 had more downtime.[4]

Before her scheduled departure time of 6:30 p.m. ship 53 was cleaned inside and the usual preflight ramp activities were in high gear. Necessary service items as well as food and drink were laid on. The two hostesses, Ginger Brisbane and Sharon Hendricks, were busy getting everything stowed and ready for the first passengers to board.

Veteran Captain Don Pauly, 47, was somewhat relaxed. He and his family—his wife, Gloria, and their four children, Donald Jr., 21; Linda, 19; Noel, 15; and Earl, 13—had just returned to their Minneapolis home a week before, having spent all of July on vacation in Europe. Pauly also had a 17-hour layover in the Crescent City from his previous flight. Captain Pauly was a 24-year veteran of commercial aviation. His pilot's log recorded 20,768 hours of flying on the job, and despite the One–11's newness, he already had 550 hours flying the type. He and his first officer, James A. Hilliker, were not Braniff originals. Each got his start with Mid-Continent, which had been swallowed up by Braniff in 1952. For the last 12 years, the two had been Braniff pilots and appreciated the additional stability and wider range of aircraft to fly and stations to visit.[5]

Prior to boarding the aircraft and bringing passengers onboard for the first short hop up to Shreveport, roughly 275 miles away, Captain Pauly had a brief conversation with Braniff station personnel in New Orleans; he would repeat the same questions to the station personnel in Kansas City later that evening. He had heard about the storm front moving down into the Omaha and Kansas City areas. At both stations, staffers noted the captain's "concern" with the weather, but neither they nor the captain thought much more about it.[6]

The first passengers of Flight 250's evening crossed the tarmac and were greeted by Sharon and Ginger, who helped them get settled. Some passengers had seen reports of Braniff's new, sexy, mod uniforms designed by Emilio Pucci. The two 21-year-old roommates, who had yet to pass their first anniversary of service with Braniff, were proud to show the spirit of the new Braniff International and its new airplanes and uniforms to everyone.

With passengers buckled up in the stylish seats covered with Herman Miller checks, stripes and solids designed by Alexander Girard, Flight 250 pushed back from the gate at Moisant field.[7] The One–11's Spey engines spooled up and reached their distinctive whining crescendo until a full-throated roar catapulted N1553 into the air out over Lake Pontchartrain. Some pilots trained in the military would compare the One–11 to a fast and nimble fighter jet like they once flew. The One–11 was 200 miles per hour faster on the routes previously served by propeller-driven aircraft, and its T-tail configuration, with the two engines in the rear, meant the ride was quieter (at least inside) as well.[8]

Flight 250's passengers relaxed as Ginger and Sharon brought them drinks and snacks. All too soon, an astonishingly short time for many who had never flown jets before, Captain Pauly was announcing the descent into the Shreveport area.

Metropolitan Airport, Kansas City, Missouri
10:55 p.m. CDT, August 6, 1966
Saturday

The other stages of 250's progress northward were just like the first; a quick, whining launch down the runway, the hostesses working against time (and the more impractical elements of Emilio Pucci's "Air Strip" uniforms) to get food, drinks, snacks and whatever else served and picked up before Captain Pauly's inevitable announcement. Sometimes the hostesses felt they were racing the One–11 itself on every stage of a flight. None of the crew and most of the passengers really paid much attention that evening to the world flying by outside their jet. Darkness in the middle of summer doesn't come very early; if they looked, they would have seen a spectacular sunset on the left side of the aircraft until almost 2200 hours.

Some passengers left the aircraft here and there, and others joined it. At Shreveport, several army soldiers who had just completed basic training at Fort Polk boarded Flight 250 for the first leg of trips home for pre-deployment leave. Most of them were headed eventually to their first tour of duty in Vietnam.[9] Turbulence was minimal and quickly over in the early stages of the flight, which was a routine and fairly mundane experience for everyone to that point. While they may have been aware that some ugly weather lay ahead between Kansas City and Omaha, there was confidence that it would be minimal and of no import.

As 250 landed at Kansas City's Metropolitan Airport, the crew could see what was lying out ahead, straddling their planned route into Omaha. It was black and ugly, pulsing with lightning and extended east and west as far as they could see. As darkness fell and the moon came out, ready for jousting with clouds and making a farm family laugh in their car, the pilots expertly guided N1553 to a smooth landing and taxied to the MKC terminal. While First Officer Hilliker did a walk-around inspection and stretched his legs, Captain Pauly went to have a talk with the weather guys.[10] He was concerned about the weather ahead. Pilots don't accumulate 21,000 hours of flight time without having a very healthy respect where the weather is concerned. And he knew that this night, a very healthy respect indeed was going to be needed, even if for just a short while.

Kansas City's weather bureau people gave him the lowdown: between Flight 250 and its Omaha destination lay numerous active thunderstorms "associated with a well-marked prefrontal squall line" (a line of active thunderstorms located in advance of a cold front paralleling the front and moving along with it, as if the front is pushing it out ahead).[11] The squall line and potentially hazardous weather was discussed in Weather Bureau Aviation Severe Weather Bulletin 447, Sigmet Bravo 3, that evening, as well as in the terminal forecast for Omaha. But Braniff's internal weather forecasts, available to the flight crew at Kansas City, did not highlight severe weather. The forecasts, observations and pilot reports from other flight operations conducted in the area were given to Captain Pauly; copies of these documents would be found in the soybean field still attached to Braniff dispatching's flight release paperwork.[12]

The flight crew didn't have only the official forecasts and Braniff's more optimistic one. Prior to 250's departure, Captain Pauly held a conversation with another Braniff crew, just arrived after a trip from Chicago. "There's a solid line of very intense thunderstorms with continuous lightning and no apparent breaks. That front is as long and mean as I've seen in a very long time. I don't think the radar reports are giving a true picture of the intensity inside those storms," that captain said. "Well, we plan to be to the west of the front line anyway, so hopefully we won't get what you guys did," Captain Pauly replied.[13]

As Pauly gathered up the paperwork for the next leg of the flight and walked out to the tarmac and the waiting N1553, 38 passengers were either already onboard or making their way out to the waiting One–11. Lightning flashed off and on in the north, but no one seemed particularly concerned. One passenger, excited by an over-stimulating day of jet travel and visiting relatives and the Kansas City Zoo, was carried onboard. Five-year-old Mitchell Kuhr went to sleep in his mother's arms in the boarding area before the aircraft arrived from Tulsa.[14] Ginger and Sharon settled everyone in as usual. Pauly and Hilliker completed preflight checks and prepared for taxi. At this point, the crew was more than ready to get home to Minneapolis.

The 42-minute trip to Omaha was planned for a cruising altitude of 20,000 feet (flight level 200). But as the One–11 taxied towards the runway, Kansas City air traffic control instructed 250 to maintain 5,000 feet after takeoff until further notice due to conflicting inbound traffic. Captain Pauly turned the snarling "Fast Back Jet" onto the piano keys of runway 19 in order to take off into the prevailing southerly winds. Departure checks completed, Flight 250 was cleared by Kansas City tower for takeoff on runway 19 and to fly the runway heading (190°—southwesterly) and to climb and maintain 5,000 feet.

Pauly set takeoff power and the Rolls-Royce Speys responded instantly, pushing the aircraft down the runway and skyward with their signature whining growl. Hilliker called out the necessary speed marks, raised the gear when Pauly called for it after a positive rate of climb was established, and retracted the flaps to create a smooth wing to slice through the sky. They were airborne at 2255 hours and quickly circled back around to a north-northwesterly heading. Immediately they could see the storm was closer and bigger than it had been when they glimpsed it earlier while landing at MKC. It was dark, boiling and suffused with pulsing plasma bursts of liquid light. The air traffic controller directed the flight to climb to the assigned flight level 200 (20,000 feet).[15]

Twelve nautical miles north on their planned route to Omaha, Flight 250 was handed off from the airport control tower to Kansas City air route traffic control (ARTC). Once Kansas City ARTC had established radar contact with 250, the crew told the controller they climbed to 6,000 feet and were already encountering a bumpy ride; they therefore preferred to remain at 5,000, rather than the much-higher planned altitude. They also passed on a request to deviate to the left of track. The controller told 250 to remain at 6,000 feet until traffic cleared at the lower level and approved the track deviation change.

At 2306, Kansas City ARTC assigned the aircraft to 5,000 feet and handed its control over to Chicago ARTC, as it was now in that center's area of responsibility.[16]

The Chicago controller greeted them, and 250's flight crew initiated a discussion on the weather displayed on the Chicago center's scopes. The controller also passed along the information that another Braniff One–11, operating southbound Flight 255 between Omaha and Kansas City, was climbing through 10,000 feet just after departure from Omaha. Flight 255 was in a position, the controller said, "to tell you what they're experiencing out ahead of you." The captain of 255 had delayed his approach into Omaha for an hour due to the turbulence; normally the two flights would not have been so close to each other.[17]

The two captains spoke by radio. Flight 255's captain reported that they were experiencing "light to moderate chop" [turbulence], which began as they reached 15 nautical miles southeast of Omaha. Flight 255 was diverting to the east and climbing to 17,000 feet and expected to be clear of turbulence after another 10 nautical miles. Captain Pauly acknowledged and thanked flight 255's crew for the information, signing off from the conversation at 23:08:30. It was to be the final transmission from Braniff International Flight 250.[18]

16

"One big, terrible nightmare": Flight 250 and Falls City

Forty-two white stakes driven into the soft earth gave silent witness today to the terror which broke over the southeastern Nebraska farmlands.
—*The Chicago Tribune*, August 8, 1966

The doctor shook his head in pouring rain and said, "There's nothing I can do."
The New York Times, August 8, 1966

7.6 miles north-northeast of Falls City, Nebraska
2:00 a.m., August 7, 1966
Sunday

In the kitchen of Tony and Vernell Schawang's (pronounced "Shwong") comfortable old farmhouse, there was coffee on the table. Tony sat and sipped and talked to visitors and Vernell kept the coffee brewing. It looked like an average scene, but it was far from it.[1] The phone was ringing continually, and the visitors were reporters, firemen and law enforcement officers. Tony and Vernell were in shock and it was 2:00 a.m. on a Sunday. Outside, across the road, a scene of unimaginable horror was being revealed in the glare of flashlights, car headlights and lightning flashes.

Tony was at the table talking with the *Falls City Journal*'s city editor, Claire Hurlbert, and its owner/publisher, Bill Schock. More reporters were trying to get through on the phone, or were already piled in cars and driving towards the farm. As Tony told Hurlbert and Schock, he and his wife and their daughter, Judy, spent an enjoyable summer Saturday evening at the home of Vernell's sister and brother-in-law, Marcelle and Bill Kuttler, not far from their own farm. The trio were returning home just after 11:00 p.m. and were a half-mile north of home when a giant ball of fire came streaking towards them. As they watched in disbelief, the remains of Braniff International Flight 250 impacted in Tony's five-acre field of soybeans, just across the road from their farmhouse.

Hurlbert recorded the family's reactions: "Though the windows of the car were down, they did not hear a thing until after they say a ball of fire that lit up the entire sky. Mrs. Schawang's first thought was that the 'whole world was on fire,' but Tony, although admittedly 'scared to death for a while,' thought it was a giant meteor and even said he gave some thought to it possibly being a satellite which may have strayed from its orbital path." Tony continued: "The only thing recognizable was one wing, the rest of the plane nothing but a giant ball of fire.... [I]t appeared that the plane spiraled end over end while falling to the ground." Tony said he slammed on the car's brakes as the One–11 came down: "There was a terrific explosion when the plane hit the ground.... [W]hat appeared to be the head end of the fuselage went back up into the air at least 500 feet before falling back to earth. Smoke mushroomed from the plane like an atom bomb."

The family's first concern was that the remains of the aircraft would hit them while they were sitting in their car. But it quickly became apparent that the jet was falling short of where they sat "petrified with fear." This was no comfort; it looked like "their house might be the final resting place." As it was, the wreckage was "only 100–150 yards from the residence and outbuildings." Overcoming his initial shock, Tony hit the gas and slid into the driveway, hurrying inside to call the fire and police departments; he also asked "the operator to summon ambulances and other emergency equipment and personnel."

Tony then ran back outside "but, with fire raging so fiercely at the time, was afraid to get any closer for fear of any possible later explosions." The caution was justified; the fires were so intense he "could have read a newspaper easily from the light at the top of the hill." Aware that such an intense fire could spread, he went back inside to the phone and called several neighbors, including Barney Schawang, Paul Schawang, Roy "Bud" Schawang and John Nutzman, who all hastened to the scene, "but the blaze remained confined to the debris—and burning bodies."

Almost immediately a "steady stream of newsmen, neighbors, friends and just plain naturally curious persons" came on the scene. Vernell found herself suddenly opening what amounted to a canteen. She served coffee and cake in the house and sent more out "into the field for firemen, police officers and volunteers." The telephone was constantly ringing "with inquiries from big-city newspapers, coast to coast and border to border." Finally, an unnamed official from Braniff's Omaha station arrived and he "tied up the phone with some urgent long distance calls that brought a well-reserved rest for the 'hosts.'"

Unfortunately, it was just the beginning. For at least two weeks more, the Schawangs' farmhouse was to be at the epicenter of a swirl of people, equipment, lights, and chaos, with airplanes and helicopters circling overhead. Hurlbert wrote, "You can take it from the Schawang family, it was an experience that will long be remembered as one big, terrible nightmare." And the "nightmare" referred not only to the crash itself, but also to everything that came afterwards.[2]

The Schawangs experienced the closest encounter with the destruction of Braniff Flight 250, but they were not the only ones to witness it. The NTSB would later identify at least 300 people all over Richardson County as eyewitnesses. One of those was Nebraska state trooper Martin Gifford. He was following a truck north of Falls City on Highway 73. He too saw the "ball of fire" in the sky to the east. He told the *Journal* that "the entire sky northeast of Falls City was alight" from the intensity of the flames. Gifford said he "pulled out around the trucker and motioned him to a stop. 'Did you just see what I saw?' the stunned trooper asked. "I sure did," the trucker replied. "What the heck?" Gifford had no idea how far away the fireball came to earth, so he "climbed on top of the truck cab in an effort to better look ... to no avail." He got back in his cruiser and headed east. He drove briefly and then the call came in: "state patrol headquarters advised by radio that an aircraft was down on the Tony Schawang farm. It was only a half-mile away." He was there in a hurry.

Gifford "did not see the burning wreckage until [he was] almost on it because it was on a hill sloping to the south and the area is extra hilly anyway. It had started to rain and [he was] somewhat subdued. A wind blew from the northwest. [He] looked around and knew there was no hope for survivors," he told Schock.[3] The scene was unchanged an hour or so later when Schock himself arrived. "The rain was accompanied by a ... storm and the east-west road was clogged.... People were standing along the fence line in bunches, looking out over the smoking wreckage. They were not saying much. The magnitude of the tragedy was difficult to comprehend." Small spots of fire were still burning here and there; the fire department was "playing water over what remained of the twin-jet airliner. Spectators were getting soaked to the skin by the spray, but they were paying it little mind."

Trooper Gifford left the fence line to count the bodies. Gifford was accompanied by Schock and Hurlbert, as well as local first responders: rural fire department chief Dale Purvis, Richardson County deputy sheriff Lanthan Camblin and Falls City chief of police Blaine Sailors. It was a scene Schock described as "apocalyptic." It was after midnight and rain fell intermittently

16. "One big, terrible nightmare"   137

with flashes of lightning, and it was undoubtedly traumatic for everyone involved.

"We moved down the hill into the soybean field, now muddy, and wet, and reached the cockpit section of the aircraft. It was a mass of twisted metal. And mangled bodies," Schock wrote. This main section of the debris field was 50 yards south from the east-west road. "It was unbelievably intact when considering the explosion which must have occurred when it met the ground. The plants in the immediate area were wilted by the heat and flames," he added. At the east side of the main wreckage field was the cockpit roof, which looked like it had been sheared off. Captain Pauly's uniform coat could be seen lying on top. The cockpit roof was still attached to a crumpled section of the forward fuselage, including the bent frame of the front left entrance. The door was lying nearby. It was in this area that searchers found the bottom half of the cockpit itself. They saw the mangled seats and front instrument panel; they also saw the bodies of the two pilots.[4]

Almost fifty years later, Bill Schock continues to carry around the image of that sight. A man was sitting slumped in a seat and his left hand was clearly grasping the flight control yoke. It was especially memorable because there was a shining gold wedding band on the third finger. Seated as he was on the right-hand side, it was obviously the body of First Officer Jim Hilliker. Schock also saw the body of one of the two Braniff hostesses onboard. She was wearing one of the new Pucci uniforms and was "lying there looking like she was asleep. Not a mark on her."[5]

As far as is known, no notes or other records identified which of the two hostesses, Sharon Hendricks or Ginger Brisbane, was sitting in the front of the aircraft for the Kansas City-Omaha stage of Flight 250. Typically, one hostess sat in the front on a jumpseat against the cockpit wall, facing backwards, while the second sat at the back bulkhead jumpseat looking forward. This way, the cabin crew could have a clear view of the passengers during take-off and landing or during turbulence. Since they would sometimes trade off sections of the cabin from stage to stage, if nothing else to relieve the boredom, which hostess was in which position was not known.

It's probable that the CAB, reconstituted as the National Transportation Safety Board (NTSB) in 1967, before the Braniff 250 investigation was finalized, initially possessed records on where each passenger and crew member was found. At the crash site, the position where each was found was marked by a white stake driven into the mud. But in response to a Freedom of Information Act request in 2012, I was informed that "the docket has been destroyed," meaning all the records, documents and testimony from the investigation had been shredded at some point in the last 46 years. Which of the two Flight 250 hostesses Bill Schock saw that night is therefore uncertain.[6]

As they walked through the debris field, Trooper Gifford made "a flashlight count of the bodies, most of which had been thrown to the south of the aircraft. We stepped gingerly as he shone his light. Bodies were everywhere and most of them were naked, or partly so. The flames had seen to that," Schock wrote. He added, "It was a grisly, grotesque sight! Occasionally a bolt of lightning momentarily lighted up the hillside and revealed more of the revolting scene. Gifford kept at his sickening task and in the rainy darkness he and his helpers counted 37 bodies."[7]

More and more outside help arrived as the night wore on. Schock and Hurlbert took pictures, which would be needed not only for the *Journal*, but also for the investigators, the airline, the manufacturer, the FBI, and so on, as well as the wire services, United Press International and the Associated Press. The Rev. John McCabe, priest of Falls City's Sts. Peter and Paul Catholic Church, was also on the scene. Vernell Schawang told reporters, "I could see him walking through the smoke, stopping every once in a while to give the last rites of the church."[8]

Officials notified as many area funeral homes as possible, knowing they would need a small army of morticians. On the scene before sunup were Bill Hodgens and Paul Sharrar (Chaney-Hodgen-Sharrar); Russell Dorr and Rodney Knaup (Dorr-Philpot); representatives of the Reavis-Macomber funeral home; and Jim Colbert and Duane McKnight (May and Timm).[9] Dr. L.V. Brennan of Falls City (and his son Richard, a future orthopedic surgeon) and Dr. A.P. Stappenbeck of Humboldt were on hand. Dr. Brennan "shook his head in pouring rain and said, 'there's nothing I can do.'" Richard Brennan was one of the area residents who saw the aircraft fall in flames, "spinning like a top. 'It kept going around and around and down and down,' he said." Although there was unfortunately nothing the physicians could do for Flight 250's victims, the crash site was extremely hazardous to anyone working. It was best to be prepared.[10]

Just after 6:00 a.m., two trucks from the local National Guard unit moved up next to the wreckage, and the assembled morticians loaded nine of the most intact bodies, six in one truck, three in the other. "The recovery operation then ceased while the morticians awaited the arrival of disposal pouches from Topeka into which the remaining bodies could be placed. Identification would

Opposite, bottom: **Another *Falls City Journal* photograph taken just hours after the crash of Braniff Flight 250. This more direct view of the roof of the cockpit area of Ship 53 shows the front windows on the left and the front exit door at right, the frame bent down and around. The dark black clothing lying in the lower center is Captain Pauly's uniform jacket, with the four stripes on the sleeve visible. It had been hanging in the back of the cockpit; most pilots do not wear uniform jackets while flying. A uniform cap is slightly lower and to the left of the jacket (*Falls City Journal*).**

16. "One big, terrible nightmare"

One of the first photographs of the remains of Braniff's BAC One–11 N1553 was taken by Bill Schock in the early morning darkness of August 7, 1966. At left is the roof of the cockpit; top left is the front exit door frame. At lower right are two of the passenger oxygen bottles which feed face masks in case of depressurization at altitude (*Falls City Journal*).

With daylight came a more revealing look at the wreckage. Debris in the cockpit area, as seen in the previous photograph taken hours before, has been moved somewhat by investigators (*Falls City Journal*).

be a problem," Schock wrote. Given the mid-air disintegration, subsequent fire and brutal ground impact, this was an understatement.[11]

Out on the road, and in the Schawang farmhouse, coffee and sandwiches were being served, and Mrs. Lawrence Hullman and Mrs. Elmer Rogers stationed themselves at the rear bumper of the rural fire department's truck to hand them out. The two *Journal* newsmen counted 14 safety patrolmen, as well as a host of officials from Braniff airlines, Federal Bureau of Investigation men, and representatives from the Federal Aviation Agency and the Civil Aeronautics Board.[12]

Opposite, bottom: First responders begin the work of removing victims of the crash of Braniff Flight 250. Behind the wreckage of the forward cabin area, local law enforcement conduct their search. The four round, puffy objects jutting out above the wreckage at top center is actually the tent top of a National Guard truck sent to move victims to Prichard Auditorium in Falls City (*Falls City Journal*).

The opposite side of the cockpit roof area seen in the previous two photographs. Braniff's distinctive black masking around the nose of the One-11 is visible around the cockpit windows at right, while a seat cushion with one of the new Girard striped fabrics lies at left. Debris at the bottom is from the roof of the passenger cabin, including a public address speaker. At upper left, some of the wreckage is still smoking (*Falls City Journal*).

Nebraska State Troopers, local morticians, and other officials consult among the still-smoking ruins of Braniff Flight 250 in Tony Schawang's soybean field. A sidewall of the fuselage is at the bottom (*Falls City Journal*).

Patrolman Gifford moved on to other duties, so Richardson County sheriff Ed Mahoney led Hurlbert and Schock through the debris once daylight came, to take more photos. By this time, the scene was being secured with "roadblocks on the roads leading into the crash site and only authorized personnel and newsmen were permitted to enter. There was work to be done and no one needed any added traffic problems."[13] But there *was* added traffic. The

16. "One big, terrible nightmare" 143

National Guard trucks are on hand to assist officials on the morning of August 7, 1966. Federal officials have already posted a notice to onlookers to stay away. The wreckage is still smoking at this point (*Falls City Journal*).

biggest thing to happen in Richardson County in years attracted not only official frenzy and sightseeing on the ground, but the air space above the county was busier than ever before. The *Journal* reported: "Air traffic was heavy over the scene of the airliner crash, 11 miles northeast of town, and at Falls City municipal airport Sunday. It was estimated that somewhere from 20 to 30 airplanes were at the local air field at times during the day. About the biggest planes were two official craft, which an observer described as probably in the $125,000 class. One local pilot who flew over the crash scene said it was about like the traffic over the Kansas City airport and that you really had to be on your toes. 'Several pilots made wrong turns, out of the official flight pattern for such areas, complicating the traffic problem,' he said."[14]

As the area lit up with the morning sun, officials began to see the scope of what they faced. "If the scene was eerie in the rain and darkness, it certainly was not better in the immediate post-dawn," Schock wrote. "Two air force caps were lying nearly on top of each other next to a small piece of wreckage. The bodies of their owners were somewhere else in the debris."

Officials found small parts of the aircraft along the east-west roadway,

An aerial view of the final position of Braniff Flight 250's right wing. As can be seen from the flattened crops, the wing hit the ground and slid along the ground toward the lower left (*Falls City Journal*).

but it was the discovery of portions of the tail section one-half mile to the southeast of the main crash site that provided an important first glimpse at what might have happened to Flight 250. This debris was lying in the fields and stretched along the compass heading the airliner had been flying, back in the direction of Kansas City. It was apparent that this was an extremely important find. As Schock, former pilot that he was, wrote, "It would seem reasonable to assume, that without the tail section, the aircraft was hopelessly out of control."[15]

Two days later, the *Journal* added to the list of the many who were also on the scene during the night. It gives a picture of how the area responded to such an unexpected and large-scale emergency:

> Within minutes of the crash, Civil Defense Director J.O. Richmond was alerted and immediately got in touch with the rescue squad, Henry and Frank Koenig, Wilfred (Butch) Young and Deputy Sheriff L.D. Camblin.
> In the meantime, the sheriff's office called the National Guard for trucks to

An unidentified Nebraska state trooper films the wreckage of BAC One-11 N1553. Heavily damaged by fire and missing its tailplane, this area of the plane still has firefighters' hoses draped over it. In front of the trooper is the port side, number one engine. The cowling of the Rolls-Royce Spey engine was flattened by the pancaking impact with the earth (*Falls City Journal*).

The starboard wing of BAC One-11 N1553. Most of the wing was snapped off under high loading in extreme turbulence. It fell to earth apart from the main wreckage, hit the ground and slid a few feet into the tall crops beyond. Civil Aeronautics Board investigators have written on the wing piece, indicating it fell rightside up (*Falls City Journal*).

Investigators examine two crucial pieces of Braniff's Ship 53: the tailplane. The largest portion of the aircraft's T-tail is at upper left, with the port side stabilizer still attached. The starboard stabilizer, near the two cars, is being turned over by officials (*Falls City Journal*).

> help remove the bodies. Master Sgt. Rex Jones, Sgt. Jun Noll, Staff Sgt. Glenn Murphy and Cpl. Walter Honea took two Guard trucks to the scene. Later, Sgt. Lyle Foster and Specialist Ray McKimmey took a third truck.

Other volunteers from the Guard unit helped at the auditorium: Ray McKim Jr., Jerry Gilliland, Wayne Becker, Alvin Comer and Lt. David Kuhlman.

> Cpl. Fred Allen and Spec. 4-C James Titus, Guardsmen, also assisted with the removal of the bodies from the burned-out plane.
> The first bodies removed were those in better condition. They were wrapped in plastic [bags] which were obtained from [the] Frontier factory. Later, plastic disaster bags were sent here from Forbes air force base, Topeka, Kas., for all of the bodies.[16]

A day later, the *Journal* added even more to the list: "State Troopers Bob Chab and Ed Pokorny, the telephone company workers, city crews both at the crash scene and the auditorium, city police at both spots, State Conservation Officer

16. *"One big, terrible nightmare"* 147

A closer view of the upside down T-tail of the BAC One–11. The rudder and starboard stabilizer are missing. A portion of Girard's "BI" lettering on the tail can still be seen with the United States flag below it (*Falls City Journal*).

Side view of the port side of the tail plane. Wreckage of the forward part has been dragged up closer to match the pieces together. Civil Aeronautics Board investigators have written instructions on the tail indicating where it should be cut up for interior examination (*Falls City Journal*).

Ship 53's starboard stabilizer separated from the top of the T-tail and fell some distance away. It was found upside down with several components ripped away (*Falls City Journal*).

On its way to Kansas City, the severed tailplane of Braniff Flight 250 pauses while a convoy of trucks is formed on Harlan Street in Falls City. Some of the wreckage was reassembled or cut up for further examination in an airport hanger. Three unidentified Falls City boys get a close-up look at the aircraft while two men discuss the situation. Seen in the background is the office of the *Falls City Journal* (*Falls City Journal*).

The *Journal's* Bill Schock was given the opportunity to fly over the site of the crash of Braniff Flight 250 in an army helicopter. He took the series of photographs seen here while on that trip. In this shot, the one mile-square section where debris was found is seen while looking northwest. The main crash site can barely be seen at the top center corner intersection of the east-west/north-south roadways (*Falls City Journal*/Bill Schock).

Ray Frandsen, two Missouri game wardens, J.N. Clifton and Glen McCloud, the Kansas trooper who brought the disposal bags for the bodies from Topeka, Robert Popkess and his son, Roger, Sabetha morticians."

Coordination of supplies was the job of Civil Defense director Richmond. Besides procuring the "extra stretchers needed, which were obtained from Community Hospital," he informed scores of air officials, and kept civil defense headquarters in Lincoln in the loop on behalf of Governor Frank Morrison. He also communicated with air traffic controllers in Chicago and Kansas City who needed information, as well as local volunteers in the field.[17]

It was an all-hands-on-deck emergency. And everyone from Vernell Schawang to FBI Director J. Edgar Hoover[18] pitched in to get the job done.

As mentioned earlier, workers placed white stakes where bodies were

As the helicopter flies towards the northwest, investigators, and their two automobiles, can be seen in the center examining pieces of the tailplane. The main crash site is at the road intersection at the top right, with the Schawang farm buildings among the trees to the left. The photograph gives a good idea of how far apart the pieces of the aircraft were when they fell to earth (*Falls City Journal*/Bill Schock).

removed. The reporters noted a poignant find: "Beaten into the mud at one such spot was a silver-colored wristwatch which had stopped at 11 minutes past 11. There was no watch band—just the watch proper—and it probably recorded the instant the jet airliner smashed into the field."[19]

Back in Falls City, the Prichard Auditorium, where a 4-H Fair had been planned, was instead pressed into service as a large temporary morgue. National Guard trucks made the trip from the crash site to the auditorium several times. Then what was undoubtedly one of the most unpleasant tasks of any such tragedy began: identification of the victims, and notification of their families. The *Journal* filed a report about this aspect as well, which was impossible to ignore in such a small town, since, as the paper noted, "the odor of death hung heavily over Prichard Auditorium today and exuded to a block or more away. The work of identification, which started almost immediately after the crash,

16. "One big, terrible nightmare"

Getting closer to the main crash site, more debris is spotted just below, at the bottom right of the photograph. The Schawangs' farmhouse is coming into better view, as is the main site (*Falls City Journal*/Bill Schock).

still has not been completed. A spokesman for Braniff said that at noon today [Monday, August 8, 1966], about 20 victims were positively identified and clearance from the civil aeronautics authorities for the removal of the identified bodies began at noon. Hopefully, the airways spokesman said that the work of releasing the bodies could be completed by tonight."

The work was done under the guidance of a four-man team of FBI identification experts who arrived on Sunday morning from Washington, D.C. They worked all day with local dentists, mainly using dental records and fingerprints. The work continued throughout the night. "Some close relatives have arrived but they are not being permitted to view the remains. Some close friends and business associates of the victims are being allowed into the emergency morgue in an effort to identify the dead." Braniff officials were quoted as being sure that "all of the caskets would have to be sealed because of the condition of the bodies." The auditorium was supposed to be "the scene of festivity, youth and laughter today as the site of the 4-H Fair," but it was instead

In this view, the helicopter has swung around and is now facing northeast while flying above the Schawang outbuildings. The main crash site of Braniff Flight 250 is now clearly visible, with officials' cars lining the roadways (*Falls City Journal*/Bill Schock).

a "grim somber scene with officials and undertakers moving quietly, grimly, efficiently among the tattered remnants of the airplane's crew and passengers."

Command and responsibility of the scene at Prichard was in the hands of Braniff officials from Dallas. John Kersey, vice president of customer service, was primarily in charge. Also on hand were the vice president of flight operations, Dan Hughes, and the public relations director, Jere Cox. Workers at the auditorium were provided a canteen by the Falls City Volunteer Firemen's Auxiliary, which "served sandwiches, cookies, coffee, milk and lemonade."

While the identification effort got underway inside, Richardson County attorney Henry F. Schepman spoke with the *Journal*. Schepman was also the coroner for the county. He said did not "regard it as necessary to hold an inquest. Inquests are called only to determine whether or not a death has resulted from unlawful causes." Schepman said he was required by law in such cases to sign all of the 42 death certificates. One of Braniff's officials added it was not "fully decided ... but it is more than likely that 'ruptured heart' [would]

16. "One big, terrible nightmare" 153

Looking northwest again, what's left of the crash debris lies among the soybeans. At left is a chunk of fuselage; the port wing can be seen on the right side of the debris. Officials tried to minimize further damage to Tony Schawang's soybean crop by using the same paths in and out of the field (*Falls City Journal*/Bill Schock).

be put on the death certificates as the cause of death." But there was much work left to be done before the official paperwork could be signed.[20]

Throughout the next four days, the auditorium was put to its grimmest use since being built by the Works Progress Administration during the Great Depression. The *Journal* described one way in which "the strong-stomached experts of the FBI" went about their work: they had "cigars gritted between their teeth for obvious reasons." Beginning the afternoon after the crash, all the dentists who had been called in began their work with dental charts brought or sent to Falls City by relatives of the victims; they also talked to other dentists all over the country. One Falls City dentist, Dr. Bob Hoban, would always remember the ruthless efficiency and "thoroughness" of the FBI.

Local morticians moved from picking up bodies in Tony Schawang's soybean field to the auditorium identification and preparation process. During this period, none of them got much sleep. They needed to make arrangements

Almost directly over the main crash site, the camera picks up more detail. At left is one of the Rolls-Royce Spey engines lying next to the port wing and rear fuselage. At lower right can be seen another fuselage section with the "-aniff" part of "Braniff International" still visible (*Falls City Journal*/Bill Schock).

for shipment of bodies as soon as investigators released them. "It wasn't just like falling off a log. There were hitches, government agencies to be contacted, clearances to obtained," wrote Bill Schock. Braniff's team also worked on identification, giving information from passenger manifests and crew personnel files to the FBI. They handled "the few personal effects retrieved from the accident scene" and coordinated sending representatives to funerals around the country. Several employees attended two funerals; one ended up representing the airline at three of them in less than 24 hours. "Braniff—or the airline's insurer—footed the bill for everything from the caskets to housing and feeding relatives and close friends coming here after the crash. And did it in the best of taste, we might add," wrote the *Journal*.

Still, all of these people were professionals. They put into action plans which were long-planned and fine-tuned. Some were still working on the investigation of the crash of United Airlines Flight 227 the previous November in

16. *"One big, terrible nightmare"*   155

By August 19, 1966, all the debris had been removed from the Schawang field and only a barren spot remained, with charred ground at its center. It was now hard to envision that an airliner had crashed there (*Falls City Journal*).

Salt Lake City, Utah, which had killed 43. They were doing a professional job for professional pay. But the people around Falls City were not. They responded automatically and humanely and in the glare of international publicity.[21]

As Bill Schock noted on August 19 as the last of the professionals and the debris left town, "[A]s Falls City screamed itself onto Page 1 of every daily newspaper in the this nation on August 7 and 8 so back we are once again to being the county seat of Richardson County. Where the cooperation of the weatherman—or perhaps the lack of it—more often than not is the summer's No. 1 news story. Basking in the news limelight for a day or two, terribly tarnished as it was by the shocking loss of life, was a deep-set experience for a small town and rural community surrounding it."

Fortunately, as the *Journal* put it, "people seem to have an inbuilt facility for rising to the occasion. This community responded with one of its finest hours. Too many persons and groups are involved to single them out. Remuneration was not even considered. It simply was the case of a warm community reacting in a wonderful way to a human tragedy." Schock reported that this

The Braniff Flight 250 crash site in the Schawangs' farm field in October 2012. The rolling hills of the Nebraska prairie are easier to see after the harvest (photograph by the author).

wasn't all just patting itself on the back: "It is not always thus, a CAB official reassured us. Our people came through with flags flying. Take it from the CAB, take it from the GIs here from Ft. Riley; take it from Braniff Airways; take it from the relatives who came on heart-breaking missions. Or leaf through the thank-you letters the post office has been delivering to Mayor Dick Baker."[22]

Beginning on the Tuesday afternoon after the accident, those who were positively identified were released by officials. Some were sent on to Omaha, others back to Kansas City, still more to destinations from Fargo, North Dakota, to Guatemala. Some went home on Braniff flights originating in either city; two went via Delta Air Lines.

At 8:30 a.m. on Wednesday, August 10, the bodies of First Officer James Hilliker and hostesses Sharon Hendricks and Ginger Brisbane were placed aboard Braniff's Flight 226, another BAC One–11, at Kansas City's Metropolitan Airport. At 11:26, the aircraft landed at Minneapolis-St. Paul, and the bodies of the three crew members were taken to the individual funeral homes chosen by the families. Fifteen minutes later, Braniff's Flight 146, a Lockheed

Marker to the memory of those who died on Braniff Flight 250. Beneath the sign is a piece of the BAC One-11 unearthed from the site (photograph by the author).

Facade of Prichard Auditorium in Falls City, October 2012. Built by the Works Progress Administration during the 1930s, the auditorium served as a makeshift morgue during the effort to identify victims of the crash of Braniff Flight 250 (*Falls City Journal*).

Electra II propjet, also left Kansas City. On board was the body of Captain Donald Pauly. That plane landed at Minneapolis-St. Paul at 2:52 p.m. and Pauly's body, too, was taken to an area funeral home. Funerals followed for the four crew members over the next two days. The first passenger funeral was that of 16-year-old Charla Ward at 2:00 p.m. on August 10, at the Hillcrest Baptist Church in Omaha. The following day, Thursday, 17 of Flight 250's passengers were laid to rest, nine more on Friday, seven on Saturday. Four private services were also held during the week.[23]

Out at the Schawang farm, the remains of N1553 were gathered up over a two-week period and the largest pieces placed on flatbed trailers. The trucks convoyed the broken jetliner to Kansas City and deposited the pieces in a hangar at the airport for long-term examination.[24] With the crowds diminished and Tony Schawang's soybean field emptied, life slowly returned to normal. Extra phone lines were removed by the telephone company, hotel rooms emptied, and the field was bulldozed flat. Officials advised Schawang that he should leave the area of the crash site fallow for a few seasons. Even with the thoroughness of the salvage experts, pieces of the aircraft would be turned up in the field for years to come.

The field across the street was a bald spot for quite some time, a visual daily reminder of what was still imprinted on their minds. Vernell and Tony were indeed traumatized by the entire experience, from the first seconds of the fireball to all the people swarming their home and fields to their appearances in the national press (they were both quoted in everything from the *Beatrice Daily Sun* to the *New York Times*). The silence after everyone went away must have been deafening as well. In fact, the Schawangs would also experience loss beyond a few soybeans. Almost immediately, they experienced difficulty staying in the farmhouse. Within months, they sold the farm and moved to another several miles to the northeast. Living (especially trying to sleep) at the scene of Flight 250's demise and its aftermath was just too painful.[25]

The work on the crash now shifted to its next phase: pinpointing what brought the One–11 plummeting to the ground. "The bodies gone, the crash investigators moved in and began poking around in the wreckage in an effort to determine why Braniff Flight 250 ended in tragedy on Tony Schawang's hillside," Schock wrote on August 8. "It looks impossible. But these men are pros at this sort of thing."[26] The professional effort to do the impossible now shifted away from Falls City.

17

"Ease power back": The Investigation of Braniff Flight 250

> *Without exception every jet mishap has set new rules of flying and has demonstrated on an ever-increasing scale that the United States and other countries are woefully behind in the aeronautical technology necessary to keep abreast of increased jet flying. All this means that as more jets are brought into service and the more flights there are, the probability gap of disaster is narrowed dramatically. Never has this been more clearly demonstrated than in the crash of a Braniff airliner near Falls City, Nebraska, on August 6, 1966.*
> —Fred McClement, *Anvil of the Gods*

The storm front was barely past the area and the bright light from the descending fireball of Flight 250 was still imprinted on witnesses' eyes when telephones began ringing from Dallas to Washington to London. Well-oiled machinery swung into action. At some point, Bill Schock managed to get a few moments from Ed Slattery, public information officer for the CAB ("as cooperative a public servant as one is likely to meet,"[1] said Schock). Slattery told the *Journal* that the process "went something like this" with Braniff Flight 250 and other air accidents.

Dick Baker was "one of the CAB's four 'Go teams' in Washington organized for just such a catastrophe. The teams rotate in an alert status for periods of a week at a time. All of the members of each accident investigation team are specialists—and expert specialists—in their fields." Just 45 minutes or less after the crash, Baker's Go Team were "at their homes, asleep, when the emergency watch duty officer alerted them to the Nebraska accident. Had Baker not been at home, he would have been carrying an electronic device known as 'the bellboy' on his person which would have been triggered by the duty officer. Had the bell rung, Baker would have headed for the nearest phone—and fast," the *Journal* reported. The bellboy was an early version of a pager.

Just two hours after the crash, Baker, his 12 team experts and two professionals from the FAA gathered at Hangar 6 of Washington National Airport. By 1:45 a.m. Falls City time, the group was airborne in a 17-passenger Gulfstream. Slattery told the *Journal*, "Three hours and 20 minutes later they landed at Rosecrans airport in St. Joseph [Missouri], where the FAA had cars for their transportation to Falls City." The FAA arranged for rooms at the Hotel Stephenson, which can be seen in John Phillip Falter's *Saturday Evening Post* cover, and is known today as the Grand Weaver. But the team went first to the crash site.

In the early morning mist, officials arrived at the still-burning wreckage. "The wet darkness had not given way very long to the post-dawn haze which has a peculiar way of adding an echo to men's voices even when an effort is being made to keep them subdued," Schock wrote. "A stranger in dark glasses and carrying a clipboard walked up to the safety patrolman in command. 'I'm Dick Baker of the CAB in Washington and I am the investigator-in-charge,' he informed the trooper. And he wasn't kidding, he *was* in charge. It had to be around 6:30 a.m."[2]

After giving instructions to secure the site, investigators convened at an organizational meeting at the hotel at 9:00 a.m. "Everyone concerned with the investigation and then on the scene sat in on the session. 'Parties of interest' were invited in by the CAB. They included the FAA, Braniff Airways, British Aircraft Co., Rolls Royce engine people, the Airline Pilots association, the Flight Traffic Dispatchers association. The FBI and the postal inspectors were along as observers," the *Journal* reported. "At the height of the investigation there were 16 CAB people on the scene, plus many more at posts away who were busy on the crash. 14 FAA representatives, three FBI men, 15 from the British Aircraft Co. and Rolls Royce and two postal observers. Other airlines had top men here in observer roles." The "other airlines top men" were most likely from American and Mohawk, the other two U.S. airlines with BAC One–11s in service. The main command post was the Gold Room at the Stephenson, and the team met there for "nightly evaluation sessions."[3]

Early on, the flight data recorder and cockpit voice recorder were found in the wreckage. The CAB ordered the Gulfstream which brought the Go Team to St. Joseph to wait while the recorders were brought from the site. It was almost immediately apparent that the FDR was charred and would perhaps be of little use. The CVR, located in the nose wheel well of the One–11, also sustained heavy damage, but was not burned. Both recorders were then immediately flown back to Washington to find out what information could be saved.

Schock (and others who around town who worked with them) were impressed by the unexpected visitors. "The caliber of the men was unquestionable. They were the best brains in the civil aviation field—both government and industry. They were professionals in the strictest sense," he wrote. He described the way spokesman Slattery worked. To the veteran, it was "all in a day's work," routine, with an occasional wrinkle, but nothing he hadn't handled before—and taken in his stride:

> And riding, apparently serenely, through it all was Slattery, an old hand at air tragedies. Press associations, newspapers, radio and television stations and magazines all fired their questions at the public information chief—in person and by telephone. If he batted an eye at the continual barrage of telephone calls, he did it on the sly.
> For direct quotes, he handed the telephone to Investigator-in-Charge Baker, a former navy pilot eternally busy with the numerous phases of the investigation. No novice at plane crashes, Baker went through the routine at the United Airlines crash at Salt Lake City and the same airline's tragedy at Parisville [sic] [Parrotsville], Tenn." (United Flight 823 was a Vickers Viscount which caught on fire from an unknown source and crashed on July 9, 1964, killing all 39 aboard.)
> Such was the operation of the CAB and the FAA."[4]

The "witness group" of investigators were given a tall order; it would be their responsibility to talk to as many of the 300 witnesses to the crash as possible. The *Journal* also talked to many of them and printed some examples of their experiences. It's an interesting read—and not just for the different perspectives of the accident. It also preserves a short peek at small-town American life on a Saturday night, such as this: "Falls City youths Kenny Brown and Mark James were talking to Cheryl Marmet and Christine Brightwell at the Breezy Hill Drive-in theater when the plane went over. Kenny said they saw what appeared to be flares being dropped from the plane." Others were watching the storm, as mentioned before, or driving home.

> Dennis Bauman, who lives three miles southeast of the site, saw the plane go over and said its lights went out before beginning the fatal descent.
> Elmer Becker, who also farms about four miles southeast of the Schawangs, said the plane was on fire as it went over his place.
> Frank Witt saw the light from the explosion at Forest City, Mo. Two other spectators said they witnessed the flash while driving between Horton and Hiawatha [Kansas].
> Alton Hinkle, who lives a half-mile west and a half-mile south of Fortescue, said the plane went over his place and appeared to be in no trouble, but his son, Wendell, saw the light from the crash minutes later.
> Bob Joy, Falls City, was returning from Rulo when he saw a big flash of light-

ning, then saw another bigger flash far north when the plane apparently hit the ground, bursting into flame.

J.N. Clifton and Glen McCloud, Missouri conservation officers, were north of Craig when the plane went down. They thought for some time it had crashed somewhere in the Missouri river bluffs.[5]

Witnesses were useful in creating a picture or discovering patterns. But even more useful would be mechanical witnesses who weren't subjective but would faithfully and accurately record minute details. Within a day of the two recovered devices being examined back in Washington, the CAB knew that one of their mechanical witnesses could provide no information; the fire destroyed Flight 250's FDR. But the CVR yielded some data, even though it was "in a rather battered condition," Slattery said.[6] It was at least something, especially considering it was only a short one month since CVRs were required in airliners.

The airlines resisted the installation of recorders on their aircraft. It was expensive, it was "Big Brother" listening to pilot talk for punitive purposes. Even though the devices had been required in Australia since the 1950s, the FAA took much longer to develop the technical requirements and regulations for American aircraft. Finally, on December 31, 1964, ten months after Northwest Flight 705 crashed in the Everglades, and having faced mounting criticism, the agency added Section 121.359—Cockpit Voice Recorders—to the federal requirements for airlines: "No certificate holder may operate a large turbine engine powered airplane or a large pressurized airplane with four reciprocating engines unless an approved cockpit voice recorder is installed in that airplane and is operated continuously from the start of the use of the checklist (before starting engines for the purpose of flight), to completion of the final checklist at the termination of the flight."[7]

The regulations were designed to take effect July 1, 1965, but were again delayed one more year. Effective July 1, 1966, CVRs were to be installed in over 2,000 airliners operating in the country. Braniff's BAC One–11s were either delivered with the recorders or they were added during maintenance downtime. Investigators on the ground in Falls City were understandably more than interested to know that N1553 was equipped with a CVR on board when it crashed. They weren't the only ones.[8]

An Illinois congressman elected in 1959, Roman C. Pucinski, made mandatory CVRs in aircraft with over six seats a cause célèbre of his political career. He championed the devices in Congress for years in spite of formidable opposition from the industry and reluctance from the FAA. It was an uphill battle. As is the usual case in situations like this, critical mass was finally reached, partly due to the losses in 1963 of Northwest 705 (February) and Pan American Flight 214 (December) over Elkton, Maryland (a Boeing 707 which was struck by lightning and exploded, killing all 81 aboard). On December 22, 1963, when the FAA announced it would be publishing its new CVR requirements within ten days, the cause of both crashes was still a mystery, especially the latter.

In a statement from his office, Congressman Pucinski announced, "If there had been a cockpit voice recorder aboard that craft [Pan Am 214], it is most likely that we would now have some idea, from the crew's conversation, whether lightning, extreme air turbulence, or some other factor was to blame. Instead we continue to fumble along in the dark in these accident investigations, often losing valuable months or even years before we learn the cause of a crash and take remedial action that may save lives in the future." He then reminded the public that the CAB still was not ready to announce a probable cause decision regarding the Northwest crash.[9] The FAA ran into trouble with the "wording of the technical requirements" as well as industry pushback (the industry charged in the *Chicago Daily News* that the agency was imposing too many controls on the airlines). But the cockpit recorder requirement could not be put off indefinitely.

Born May 3, 1919, Roman Pucinski possessed life experiences similar to Eastern's Grant Newby and Falls City's Bill Schock; he was a veteran war pilot. And like Schock, there was newspaper ink in his blood. An Illinois House of Representatives resolution passed when Pucinski died in 2002 gave the details of those experiences.[10] Pucinski, the resolution stated, was lovingly known as "Pooch" to colleagues and friends. He "grew up in a heavily Polish neighborhood that is now Wicker Park; his youth was shaped by his father's abandonment of his mother and siblings when he was a child and by the Depression in his pre-teen years when he wore government-issued shoes; he helped his mother Lidia, later a personality on a radio station he owned, support their family by selling Magic Washer soap to local grocery stores and chocolate to office workers after school."

In January 1939, Pucinski was a student at Northwestern University when he took on reporting duties for the *Chicago Times* (later known as the *Chicago Sun-Times*). He entered John Marshall Law School "but never took the bar exam because he was too busy covering the 1948 presidential election." During World War II, Pucinski, like Bill Schock, temporarily left the newspaper world and joined the U.S. Army. His flight experience came in the Pacific theater; after promotion to a captaincy, he had bombardier duties "in the first B-29 bomb raid on Tokyo in 1944 and later flew 49 bombing missions over Japan."

He returned to his job at the Chicago newspaper, making his way eventually to the Chicago City Hall beat, "a job that exposed him to opportunities in politics." He was described as "articulate and never at a loss for words." Pucinski's Polish roots and fluency in the language, as well as his contacts among the Illinois congressional delegation, resulted in a one-year assignment in Washington. He spent 1952 as a "bilingual chief investigator for a special House subcommittee investigating the Katyn Forest massacre of thousands of Polish military officers by the Soviets during the war," a stint that added to his political experience and visibility.

Chicago's legendary Mayor Richard J. Daley encouraged a subsequent run for Congress, so Pucinski "entered the 11th District race on the City's Far Northwest Side in 1956." He lost that race to the incumbent, but he tried again in 1958 and was elected. Entering the U.S. House in January of 1959, he built a reputation in a short time as "the hardest working member of the Illinois congressional delegation." U.S. Senate Majority Leader Lyndon B. Johnson, chosen to run with John Kennedy a year later, referred to Pucinski as "the workhorse." For Johnson, a legendary workaholic, to refer to anyone else as a "workhorse" was high praise.

Pucinski's congressional career ran for 14 years. In 1973, after losing a race for the U.S. Senate, he returned home and became a Chicago alderman until his retirement in 1991. During his time in the House, Pucinski "championed airline safety.... On December 18, 1998, he was cited by the ... FAA for his role as a freshman congressman" for championing the mandatory installation of CVRs in airliners. This "Silver Medal of Distinguished Service" was handed out by the agency with the industry's representatives in attendance at a ceremony in the Polish Museum of America." It was quite a reversal for an agency that had dragged its feet for years over the issue (and may have considered the "freshman from Illinois" a thorn in its side) and an industry which donated heavily to anyone who would try to remove Pucinski from Congress.

At the 1998 ceremony which honored him for his efforts, Pucinski told those assembled that the primary objection by the airliners was based on a fear that expensive litigation would result if cockpit conversations were revealed. He added, "Their pattern had always been to blame the pilots as a way to hide maintenance problems." It was a bit more complex than that, but the industry's concern, operating under wide-ranging regulation in a country largely perceived as being "lawsuit happy," was very real.[11] The arguments against CVRs were nothing new; similar opposition was initially mounted against the FDR as well. The FAA's own official history records the ups and downs of the FDR effort:

> Aug 5, 1957: The Civil Aeronautics Board adopted a rule requiring an approved Flight Data Recorder (FDR) aboard air carrier and commercial airplanes of more than 12,500 pounds maximum certificated takeoff weight, with compliance by Jul 15, 1958. The FDRs were to be capable of recording time, air speed, altitude, vertical acceleration, and heading. In adopting the rule, CAB stated that FDRs would be invaluable in investigating accidents and such incidents as

extreme vertical accelerations. (At first, the rule applied only to aircraft certificated for operations above 25,000 feet, but this limitation was dropped in an amendment issued on Jul 12, 1960.)

On two previous occasions, CAB rescinded a similar rule. Effective Apr 1, 1941, CAB required a simpler type of FDR on certain carriers; but on Jun 9, 1944, the board found that operators could not properly maintain their recorders because of wartime material shortages. On Sep 15, 1947, the board again adopted a rule requiring FDRs on aircraft in scheduled air transportation. Contrary to expectations, however, no recording device of proven reliability was readily available, and CAB rescinded the rule on Jun 30, 1948, one day before its effective date.

FDRs were finally required as of 1960 and began to prove their usefulness almost immediately.[12] The effort to mandate CVRs possessed a similar up-and-down history, but by the time the agency issued its award to Pucinski in 1998, more than 25,000 aircraft were outfitted with CVRs.

Given that Pucinski worked on the CVR issue for over six years (and that he was still in the U.S. House), his was a vested interest in the first crash to occur with one of the devices aboard: Braniff Flight 250. When the recorder arrived in Washington, it took weeks for experts to extract understandable audio from it, but they were eventually successful. The CAB invited Congressman Pucinski to listen to the results on September 12. Afterwards, Pucinski told reporters what he heard: "While it is highly speculative and dangerous to draw conclusions at this point in the investigation, it is possible that the plane might have broken up and that the pilot and copilot might have been knocked out before the crash." He added that investigators were "certain" they could hear a "swirl of wind" and the tape confirmed that the aircraft encountered "a terribly violent storm and that the pilot and co-pilot were attempting to find openings to get out of it."

Under the circumstances, it would have been understandable if Pucinski had felt a sense of vindication. As he noted, it was "the first time ... the CAB [had] the precise time of the crash because the sounds of the plane crashing can be heard on the tape." Even at that point the analysis of the tape was still not complete. Another month of work would be needed. But investigators now possessed a hugely important tool to figure out what happened over Falls City.[13]

When the CAB conducted its official hearings into the crash the following December in Omaha, the transcript of the recording was entered into the record. A few parts or words were never deciphered, but the board's audio experts worked some magic. The last words spoken by Captain Pauly and First Officer Hilliker, as well as any other sounds audible in the cockpit, could now be heard quite clearly. The full recording was 32½ minutes long, enough to let the CAB ride along in the cockpit almost from the moment of takeoff in Kansas City. The investigative team's review of the tape showed that "eight minutes after takeoff the crew requested a change in assigned altitude from

17. "Ease power back"

FL 200 [20,000 feet] to 5,000 feet. At 11:04:44 p.m. just after a short crew conversation referring to a hole in the line of clouds, Flight 250 requested permission to deviate to the left of course."[14]

Captain Pauly requested a report on the weather ahead from the en route air traffic controller in Chicago handling the flight. He told the crew at 11:06:56 that the squall line in front of them was "pretty solid all the way from west of Pawnee to Des Moines." This was followed at 11:07:18 by "intermittent cockpit conversation regarding deviation to Pawnee which ended with '... we're not that far away from it. Pawnee is a hundred twelve four if you want it.'" (112.4 was the radio frequency of the navigation VORTAC at Pawnee, Kansas.) This statement was identified as being made by First Officer Hilliker.

In the transcript, Captain Pauly is identified as voice 1 and First Officer Hilliker as voice 2. At this point there was some interaction with one of the two hostesses; whether it was Hendricks or Brisbane could not be determined.

Voice 1: "Ah—we're going to buckle you down again."
Stewardess: "Yeah, I figured as much."
Voice 2: "It's a little rough."
Stewardess: "Are we going down?" [From the tone of her voice, the CAB believed she was referring to the aircraft descending to a lower altitude, not crashing.]
Voice 1. "We think so."

The last three minutes of conversation, since they were of primary interest to investigators, were printed in newspapers nationwide to give the public a better idea of what CVRs could reveal:

11:08:39—Voice 1: "Well, flight 255 says we can go to Pawnee by heading that way.... It's clear west of Pawnee ... but Pawnee that's no good ... well it's a possibility ... [unintelligible]"
11:08:44—Voice 2: "What! ... This looks pretty ... ah ... ah ... bad to me."
11:09:52—[Sound of landing gear horn blowing, which indicated that the crew had pulled the throttles back to reduce power; the horn normally blows in such a case.]
11:10:24—Voice 1: "We got to ... ah ... ah ... get some [unintelligible]."
Voice 2: "Heading 259. Thirty six, 37 DME from Pawnee."
Voice 1: "Huh."
11:11:20—Voice 2: "[unintelligible] ... one, twelve, four. [unintelligible] one, sixteen, three."
11:11:30— "One [unintelligible sound uttered]."
Voice 1: "Flaps."
Voice 2: ["At" or "eight"] Zero.
Voice 1: "Ease power back ... [unintelligible]"
11:11:42—[Sound similar to rushing air begins, increasing in intensity.]
11:11:50—[First loud sound indicative of structural breakup occurs, followed by air noises, speaker transmissions apparently coming in on radio and horns blowing.]

The tape terminates at 11:12:08 with a tremendous crash/ground impact. The tape time supports eyewitness reports.[15] As the transcript shows, after the last words of the crew ("ease power back"), there was a sudden noise described as "rushing air" that increased in volume, said the CAB, within 0.16 seconds. This occurred at 23:11:42, which corresponded to eyewitness accounts, not to mention the stopped watch found in the debris by the *Journal*'s reporters. This was followed within eight seconds by an unidentified sound, then an "electronic flutter sound followed by four klaxon horn sounds" (indicating that the stall protection system activated, as well as the stick pusher, which attempts to lower the nose in the event of a stall.)

According to the CVR, the time from the sudden "rushing air" noise to impact with the ground was just 26 seconds. That there was no further sound from the crew in those 26 seconds was another important clue to the CAB. In a two-man crew, especially in an emergency, communication is constant. As would be shown in recordings of later accidents, crew members react volubly in the last seconds, unless they are already unconscious or dead. It was clear to investigators that Pauly and Hilliker were one or the other from the split-second onset of the "rushing air" noise. Flight 250 must have been hit by something unimaginably powerful—so powerful, in fact, that it was considered an impossibility.[16]

Representatives (and rank-and-file mechanics) of both Braniff and BAC believed that the One-11 was constructed so well that only a very serious bomb could even rupture the skin. In fact, BAC took special pride in its sale literature that the jetliner was designed and built to strengths far beyond that required by regulators in both the United Kingdom and the United States. But in the absence of any evidence of explosive residues or other indicators of a bomb, both investigators and stake-holders like BAC were faced with a disturbing thought: either something went wrong with N1553 itself (perhaps as with both the de Haviland Comet and the Lockheed Electra) or it wasn't as strong as they thought.[17] The idea that an external force was powerful enough to upset a jet was already proven, as discussed here. In this case, it was being suggested that such a force was also powerful enough to break apart a jet *before* the aerodynamic overloading of a high-speed dive could do so. It was what Najeeb Halaby and all the others cited in the *Life* article of 1964 were fearing: that "hidden giant in the sky."[18]

While the official investigation went on, British Aircraft decided the industry needed to know more about that giant in the sky. There had been enough foot-dragging (just as there was with the recorders mandate). They found a man who was rapidly building a reputation for being able to unmask multiple sky giants and he was hired to conduct his own investigation and testify in the official hearing before the board. His name was Ted Fujita.

18

Mr. Tornado:
Dr. Ted Fujita and the Squall Line of August 6–7, 1966

In order to avoid possible squall-line turbulence, the author recommends that the penetration of squall lines just ahead of active cells be avoided. If such a penetration is absolutely necessary, the pilot should try to climb slightly above the cloud base during the penetration or else penetrate the line at least 25 to 30 miles away from the edge of solid echoes in mature downdraft stages.
—Dr. Ted Fujita, November 1966

Every airliner accident has its "it could have been me" stories. A premonition, a missed connection, a last-minute change in plans—the reasons vary, but the storytelling has the same theme: "I would be dead right now if it weren't for that traffic jam," and so on. There is usually no speculation on what circumstances came together to put the victims on the doomed aircraft, but that would put a bit of a damper on the story.

Two days after the crash of Braniff Flight 250, one of these stories made not only the *Falls City Journal*, but also the UPI wire service. An aide to U.S. congressman Clair Callan (D-Lincoln) told reporters that the congressman barely escaped death on the flight because he stayed an extra day in Washington for hearings on a rural electrification bill. Congressman Callan usually took other airlines in his commuting between Nebraska and Washington. But a nationwide airline strike by 35,000 workers, underway since July 8, affected United, Northwest, National, TWA and Eastern and shut down 60 percent of the industry. This meant that Callan and many others scrambled to find seats on other carriers. During the 43 days of the strike, Callan commuted on

Braniff via Dallas and Kansas City. He used Flight 250 several times between Kansas City and Omaha. If it had not been for the electrification bill hearings, he would have been aboard the flight once again the night of August 6, since he was scheduled to make an appearance at the Czech Festival in Wilber Sunday morning.[1]

"It could have been me" stories are especially abundant after disasters much larger than plane crashes. Such is the case with the nuclear bombings of Hiroshima and Nagasaki, Japan, in 1945. For the rest of his life, one man would tell the story of how he avoided not just one, but both bombings. In 1939, the man's father insisted that his son attend Meiji College (now the Kyushu Institute of Technology) near his home instead of the son's first choice, a teachers' college in Hiroshima. Since his father died shortly afterwards, the son honored his elder's wishes. He would always believe that his father's choice saved him from Hiroshima's nuclear holocaust. The man's "It could have been me" story for Nagasaki was different. On that fateful day, Kokura was actually the primary target, but it was fogged in that day; the B-29 "Bockscar" dropped its bomb, "Fat Man," on Nagasaki instead. The man was living in Kokura and so was saved by the fog, fortunately for the future.[2]

The man's name? Tetsuya "Ted" Fujita.

Before he became famous for his work ethic, remarkable data gathering and analysis skills, his work on microbursts (so deadly to airliners) and the nickname "Mr. Tornado," Ted Fujita built a careful foundation in Japan for his future work. He was born in Kitakyushu City on October 23, 1920. His father, Tomojiro, was a grammar school teacher who named his son by combining what he said was the Chinese character *Tetsu*, meaning philosophy, and *ya*, a Japanese suffix to a boy's name. His father died in 1939, his mother, Yoshie, in 1941.

A story often told to illustrate Fujita's analytical mind involved a 1936 school trip to the Yabakei Gorge not far from Kitakyushu City. According to legend, a Buddhist monk named Zenkai chose to dig a tunnel in the rocky gorge because he noticed that travelers endured hardships in just negotiating particularly dangerous spots. He dug what is known as the *Aono Domon* (Blue Tunnel) using only a hammer and chisel. The effort took 30 years.

When a teacher asked the young Tetsuya Fujita if he admired this accomplishment, his reply was unexpected. He told his teacher that he would spend the first 15 years inventing a "digging machine" then spend the last half digging. Therefore, the tunnel could be finished in only 15 years after the invention; therefore the 30 years of effort would result in both a new, useful machine and a safer tunnel for travelers. His teacher, apparently unamused, said Fujita failed to appreciate the monk's spiritual attainment and gave him a failing grade.[3]

In spite of this momentary setback, Fujita's schooling proceeded otherwise normally and he graduated in March 1939, two months before his father's death, and from Meiji College with a degree in mechanical engineering. He embarked on postgraduate work as a faculty member of the college in 1943 and worked on defense contracts from the Imperial Japanese Navy. In July 1944, he worked with Japanese marines at Saipan to generate nomograms (graphic representations of mathematical relationships used by engineers for fast, precise calculations of complicated formulas). In this application, Fujita was creating a graphic picture of enemy aircraft positions via three-dimensional triangulation to give a better picture of bombing raids.

Fujita was in Tokyo in March 1945 when the largest raids of the war took place, including the incendiary raid of March 9–10, which devastated 17 square miles of the city, destroyed a quarter-million homes and, by some estimates, killed between 100,000 and 300,000 people—more than would be killed at Hiroshima and Nagasaki together. He escaped injury, but found unexploded bombs in the neighborhood.[4] In September, Fujita was again contracted by the navy to perform damage surveys on the two atom-bombed cities, located (by his calculations) approximately 135 miles (217 or so km) from his Kitakyushu home in the Tobata district.

He went to Nagasaki first to witness shock wave effects on trees and structures. He and a group of his students conducted the surveys on foot. The surveys would prove to be crucial learning for the future. Fujita saw that trees in a radius around the epicenters of the bomb blasts in both cities were snapped off and fell horizontally away from Ground Zero. He noticed this very large starburst pattern, which was caused by the bomb's force going straight down from the midair detonation point, hitting the ground, and radiating outward. This "downburst" pattern would feature prominently in his professional activities for the rest of his life. The group took what precautions they could against radiation aftereffects, but they still experienced diarrhea, nausea and headaches. Fujita told them not to sit on or touch anything in the blast area. He himself was not, he said, adversely affected over the rest of his life.[5]

Soon after the war, Fujita's focus turned to the weather and other natural phenomena. As a young student he constructed telescopes out of cardboard tubes and charted sunspots. As an adult, he became increasingly interested in the weather on earth. In August of 1947, he collected reports and recording traces from 30 weather stations in western Japan to build a picture of storm development. Fujita wrote of seeing a thunderstorm near Kitakyushu in July 1948. When he heard the approaching thunder, he took a pencil and paper and began recording the direction of lightning, as well as the time between the flash and the subsequent thunderclap. For an hour and a half, he watched and recorded the storm and its 33 lightning strikes as it moved from left to right in front of him. Fujita plotted the motion vector of the storm and created a time-scale conversion method to classify lightning strikes in three different types of size and position. In September, he observed damage from a rare tornado on Kyushu, his first.[6] It was a method that he would use continuously. Even after his retirement, he meticulously plotted data from a variety of sources, such as his home mortgage, investments and retirement credits, as well as his blood pressure and weight. He also plotted the location and weather data from a trip on Concorde in 1989.[7] Fujita's gift of data analysis would take him all over the world, but it was in the United States where it would be needed most.

During his childhood, he was taught that four fearful things were earthquake, lightning, fire and Father. In the midwestern United States, he said, an equivalent fear was tornado, lightning, fire and crime. According to his reckoning, in the 75 years between 1916 and 1991, 11,944 deaths due to tornadoes were recorded. Fujita conducted lectures for the Japanese weather services on his collection and analysis methods, and decided to publish the findings in English-language journals. Reading in these journals about the work of the University of Chicago professor Horace Byers, director of the Thunderstorm Project in the United States, Fujita began a correspondence which resulted in an invitation to lecture in 1953 and to join the faculty in 1955. For the next decade, he worked hard on collection, analysis, formulation of theories and publication of the results, building a solid understanding of the often deadly weather of the U.S.[8]

By the time Braniff Flight 250 went down in 1966, Dr. Fujita possessed an international reputation in understanding thunderstorms, tornadoes and the forces contained within them. British Aircraft Corporation decided he was the expert best qualified to analyze what happened in the air over Falls City. While he had never analyzed an aircraft accident prior to the 1966 BAC contract, it was not a handicap. From Hiroshima and Nagasaki onward, Dr. Fujita did have extensive experience with analyzing damage done to both natural and man-made structures. An aircraft is just such a structure, albeit one that moves through the air. By the time he was given the contract, investigators could tell, from finding N1553's tail structure, part of the right wing, and other pieces in a line extending as much as a mile-long, that the aircraft broke up in mid-air. How this could have been possible was the main focus of both the CAB and Dr. Fujita.

The latter's analysis was discussed in a three-part report written in November 1966 and given to BAC and the official investigation team. It was made part of the board's public documents when Dr. Fujita testified at the public hearings into the crash in December 1966 in Omaha.[9] He gave his report a typical scientific title: "Detailed Investigation of Mesometeorological

Conditions of the Squall Line of August 6–7, 1966, Which Crossed the Air Route Between Kansas City, Missouri and Omaha, Nebraska." Each part was around 30 pages long. Part I contained text and figures, Part II was full of references and reproductions, and Part III was a discussion: "Turbulence in Relation to the Squall Line." (The CAB defined "squall line" at the time as "a line or narrow band of active thunderstorms located frequently in advance of a cold front, usually oriented roughly parallel to the cold front and moving in generally the same manner as that front.")

In part I, Dr. Fujita described how he used data collected by weather stations in the entire Falls City region, such as Maryville, Missouri, some 60 miles away as the crow flies. Such data included rainfall rates and amounts, wind speeds, barometric pressure, humidity, and so on. Given his expertise in analysis, he was able to create a complete picture of the moment that Flight 250 broke up.[10] This picture showed several things, Dr. Fujita wrote. First, barometric pressures were medium to strong. Second, precipitation "averaged over a large area of the squall line was light to medium," but variations in amounts between nearby stations were "unusually large." Third, there was a cold dome over the area, and a wind-shift line stretched through from west-southwest up to east-northeast. Winds at the 6,000-feet-and-above levels were from the northwest. And last, this line could be "precisely determined by a radar thin line." There were areas of very "heavy rain behind the wind-shift line."[11]

The data picture thus revealed to Dr. Fujita, as if in a photo, clearly what was going on as Flight 250 flew over Richardson County. "It became evident that the center of rain ... was located about 30 miles to the northeast of the accident site. A total of 1.58 inches of rain fell at Maryville [recording site] 2E; while a recording gauge at [site] M-1, [seven miles north-northwest of] Maryville revealed that 0.75 and 1.00 inch fell, respectively, within 30 min and 45 min after the rain began shortly before [11:00 p.m. CST]," Dr. Fujita wrote. Thus, the heaviest rainfall was located across the Missouri River 12 minutes before the accident.

Dr. Fujita reviewed wind speed reports from 25 locations and converted them "into wind scales 2 through 7," with scale 2 representing gusts of 30 mph or less and scale 7 those over 70 mph. "When these estimated wind scales ... were plotted ... it was found that the wind speed increased eastward from about a 2 to a 7 scale. Such an increase concurs with the expectation that the strongest outflow should be seen to the southeast of the heavy rain."[12] He used his own theoretical calculations derived over the previous 20 years to show that "the amount of excess pressure produced by a precipitation system is proportional to the amount of surface rain if the height of the convective cloud base remains unchanged. Since the rainfall amount to the west of the accident site was about 0.25 inches, while that to the east was over 1.00 inch, we may expect that the rain-induced cold air mass to the east was about four times larger that that to the west of the [crash] site." All of this basically added up to strong winds aloft.

Dr. Fujita's analysis of the aircraft's position in time and space showed that "the aircraft crossed the wind-shift line where the horizontal wind shear was the strongest, at levels between 2000 and 3000 feet above the ground." He concluded that "the aircraft crossed above the surface wind-shift line at [11:10 p.m.] CST. The time, the location, and the altitude of the aircraft after [11:10 p.m.] were most favorable to the development of roll circulations with horizontal vortex axes parallel to the wind-shift line and for that of circulations with vertical vortex axes."[13] The rainfall and wind speed data revealed to Dr. Fujita, in other words, that Braniff Flight 250 passed through the approaching squall line in its path where conditions "were most favorable" for rolling horizontal and circulating vertical winds. The aircraft flew into a maelstrom of high-speed wind swirling parallel and perpendicular to the ground. It was confirmation that forces aloft were indeed extremely strong. But were they strong enough to bring down a nearly new BAC One-11?

In part II of the report, "References and Reproductions," Dr. Fujita discussed the rationale for his conclusions, based on the previous decades of research—his and others. He made a meticulous case for his findings.[14] For instance, he noted that previous research revealed that "surface pressure rises and oscillates appreciably when thunderstorms pass over a station." It was postulated as early as 1935 that "the cold air contributing to the rise is produced by precipitation."[15]

The published results of the Thunderstorm Project of 1949 (which was itself instrumental in bringing Fujita to the U.S.) "revealed the fact that the downdrafts descending through the storm clouds are the source of the air forming the cold dome over the ground." Dr. Horace Byers himself "pointed out and explained two important features of thunderstorm pressure traces, the 'dome' and the 'nose.'" And the Thunderstorm Project also proved that "average cloud bases was between 1,000 and 5,000 feet."

There were also studies which confirmed the "existence of pressure-jump waves propagating as far as several hundred miles without affecting the surface temperature field, as well as the fact that the velocity of the line usually differs appreciably from the upper wind velocity at any level up to the Tropopause." And further work—in 1956 and again in 1958—proved the frequent appearance of a mesoscale dome-type pressure disturbance accompanied by squall-line thunderstorms."[16] All of this "mesometeorology speak" added up to building a foundation over thirty pages for Dr. Fujita's case that some extreme weather existed in the Falls City area the night of August 6.

The final report of the series was used by Dr. Fujita to "establish the relationship between the rain area and the flight paths through detailed analysis" in order to show the "extent of turbulence in relation to the squall line."[17] After he completed two parts of his report, Dr. Fujita used the "every-six-second radar pictures from Omaha center" to plot positions of Braniff Flights 250 (BN250) and 255 (BN255), as well as the flight data recorder report pulled from Frontier Airlines Flight 564 (FL564). Both BN255 and FL564 penetrated the squall line within 20 miles of the Maryville rain area detailed in the first part of the document.[18]

Dr. Fujita relied on all the techniques he developed over the years. He applied mathematical formulas similar to the nomographs he created for the Japanese navy during the war to give a visualization of the locations and paths taken by all three airliners, as well as the position of the wind shear front and areas of heavy rain. The resulting figures allowed him to "speculate that the winds just behind the line ... were between 35 and 40 kt." and that this speed was "1.5 to 1.8 times faster" than the speed the line itself was moving to the southeast. Behind this area of higher velocities (30 miles northwest of the squall line), the winds slowed back down from 40 to 30 knots.[19]

Where in 1944 Dr. Fujita plotted enemy aircraft positions by hand, in 1966 some early computing power became available. The University of Chicago's own IBM 7094 mainframe computer was used to speed up the process and make it more accurate. The computer allowed accurate plots of aircraft positions at ten-second intervals.[20] The resulting graphic output showed that "FL564 [a Convair 580 twin-prop] passed to the west of the Maryville rain area at 2,500 feet, while BN255 [another BAC One–11] flew to the east of the rain area at 7,000 ft from the cold to the warm sector. BN250 headed northwest at 5,000 ft." The computed positions of FL564 "showed that the recorded severe turbulence between [10:52:30 and 10:54:30 p.m. CST] started upon emergence from precipitation areas and ended upon crossing the wind-shift line."[21]

Frontier 564's captain stated, "Three minutes later we were in the clear and started our climb to 5,000 ft. At this time, the turbulence intensified and we encountered severe, heavy gusts throwing the pillows and blankets from the over-head storage.... The recorded vertical acceleration from the FDR was +2.85 G." As bad as the ride was, Frontier 564 was fortunate not to be in the worst part of the system. The flight path and the turbulence encountered by FL564 were superimposed upon a radar picture taken at 10:52 p.m. The flight path was probably very close to the west edge of the cold outflow from two areas of heavy rain labeled as M1 and M2. "Fortunately, however, the aircraft got through this outflow region when the outflow was in the early stage of development."

The next sentence is significant in that it contains an early reference to something Dr. Fujita and the meteorology community would study, debate and even argue about for the next two decades: "The quick growth and amalgamation of [the two rainfall areas M1 and M2] suggest that these cells have started producing significant *downdraft* which quickly spreads out from these echo areas" (emphasis added). Dr. Fujita was already seeing the same kind of effect in thunderstorms as he did in the two nuclear bomb blasts, and was describing it with the early term, "downdraft."[22]

Further analysis of the path computations showed that "detailed investigation of the FL564 and BN255 paths in relation to echoes appearing in the [Des Moines, IA] radar pictures taken at about six-minute intervals revealed that the region of severe turbulence was located in the clear region ahead of the precipitation area just behind the wind-shift line where the cold-air outflow replaced the pre-existing warm, moist air," Dr. Fujita wrote. This "turbulent layer extended to about 7,000 feet … the approximate height of the cloud base." The three aircraft penetrated the line at 2,500, 5,000 and 7,000 feet within 20 nautical miles from the echo boundary of the Maryville rain introduced in part one of this report. "The times of penetration were between [10:54 and 11:22 p.m. CST. Even though the penetrations took place within short distances from each other, the time and location of each penetration differed greatly, due to rapid changes" in the thunderstorm complex.[23] Dr. Fujita described the minute details of each of the three flights this way:

> FL564, 1,500 feet, ¼ of the distance from the ground to the cloud base, at [11:54 p.m. CST], Maryville rain areas were about to organize into a solid echo with developing downdraft; 2.85 Gs [of turbulence force] encountered on crossing the line. [This flight reached its St. Joseph, Missouri, destination safely.]
>
> BN250, 4,000 feet above ground, about ⅔ of the distance between the ground to the cloud base, at [11:11 p.m. CST] when Maryville rain and accompanying downdraft were most intense; Both roll and column circulations were expected at the location of the penetration. [This is the accident flight.]
>
> BN255, at 6,000 ft, just below the cloud base, [11:11 p.m. CST] when Maryville rain was still intense but in early stages of decay. Moderate to severe turbulence encountered after emerging from rain area.[24] [This flight reached its Kansas City destination safely.]

The 90-page, three-part report therefore meticulously created a state-of-the-art look at the weather picture at the time of Braniff Flight 250's downing. As mentioned, Dr. Fujita testified about his findings at the official public hearing a month after producing the report. Dr. Fujita finished his work with some sage advice, from a professional weather scientist, to professional pilots: "In order to avoid possible squall-line turbulence, the author recommends that the penetration of squall lines just ahead of active cells be avoided. If such a penetration is absolutely necessary, the pilot should try to climb slightly above

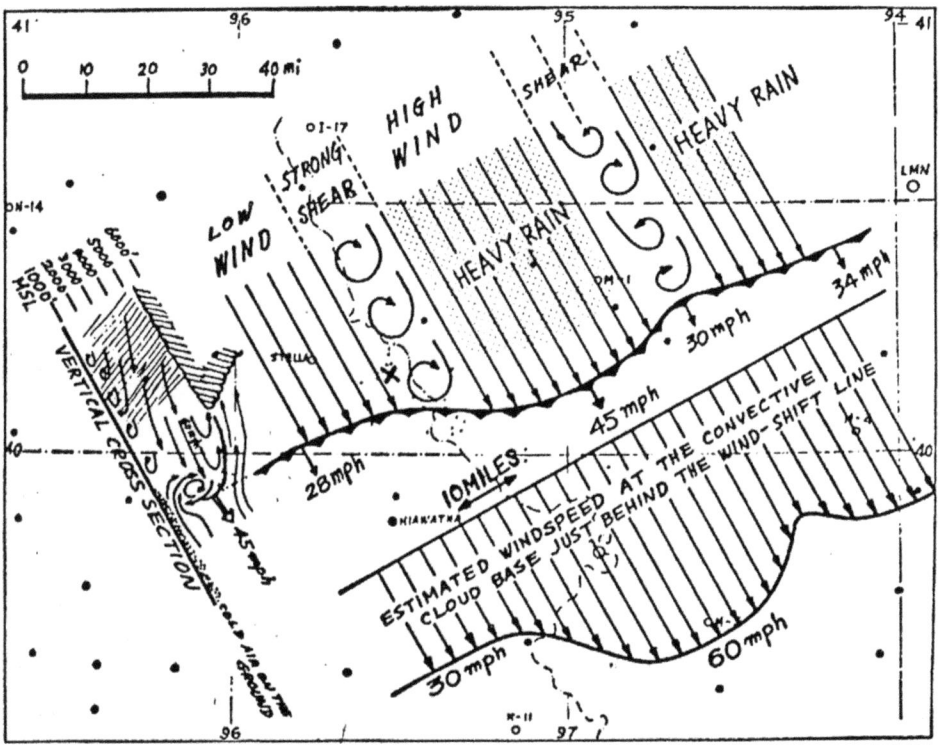

Dr. Tetsuya Fujita's graphic of the squall line which brought down Braniff Flight 250 accompanied his testimony at the Civil Aeronautics Board hearing into the crash in December 1966. It is one of the few exhibits which survived the public docket's destruction. The crash site is marked by an X at left center. The drawing at left is a vertical cross section better viewed by holding the graphic on its side. An example of Dr. Fujita's meticulous attention to detail, the drawing shows that the flight went down at one of the squall line's most turbulent areas (Civil Aeronautics Board / National Transportation Safety Board).

the cloud base during the penetration or else penetrate the line at least 25 to 30 miles away from the edge of solid echoes in mature downdraft stages."[25]

It was one more important step in understanding how to avoid jet upsets or crashes in turbulence. But just how this proven intense thunderstorm activity resulted in Braniff Flight 250's in-flight destruction was the open question. Thanks to Dr. Fujita, both BAC and the CAB now possessed a clear picture of the weather; now they needed a clear picture of what that weather had done to N1553. The airline industry and regulatory agencies moved towards a better understanding of how turbulent forces act on aircraft, while Dr. Fujita, now thoroughly Americanized with the nickname "Ted," headed off to a long quest to better understand the forces themselves.

The researcher's experience with the Flight 250 investigation would be repeated, several times: in 1975 with Continental Airlines Flight 426 in Denver and Eastern Air Lines Flight 66 in New York City; in 1979 with Pan American Flight 759 in New Orleans; in 1985 with Delta Air Lines Flight 191 in Dallas; and in 1994 with USAir Flight 1016 in Charlotte. Dr. Fujita either researched, or assisted in the investigations of, all these accidents. More important, all the massive amount of his research stretching from World War II to his retirement in the 1990s resulted in the development of his "Fujita Scale of Tornado Intensity." The F-Scale was used between 1971 and 2007 to describe tornado damage, especially in the United States, in ways understandable to both researchers and the public. It was modified in 2007, but Tetsuya Fujita's name is still attached: The Enhanced Fujita Scale now rates tornadoes in terms of the damage they produce, with six categories from EF-0 (winds of 65–85 mph) to EF-5 (winds greater than 200 mph).[26]

The long study by "Mr. Tornado" of complex, damaging wind systems took in atomic bomb blasts, thunderstorms, tornadoes—and airliner accidents—providing important pieces to deadly puzzles.

As for the deadly puzzle that was Braniff Flight 250, the CAB was ready to fit pieces together. With the investigative team's own work as well as Dr. Fujita's testimony at its official hearing on the crash in December 1966, the team started combining the various strands of their work together into a clearer picture. They could now nail down some solid answers.

19

Knocked from the Sky: A High-Intensity Force

The probable cause of this accident was inflight structural failure caused by extreme turbulence during operation of the aircraft in an area of avoidable hazardous weather.
—National Transportation Safety Board, April 18, 1968

While the weather picture was being investigated, minute examination of the aircraft was underway. There were metallurgical studies to be performed, chemical analyses to be made. Why did the One–11 come apart during turbulence that other flights traversed with no result other than an unpleasant ride? The key was in both the storm and the fact that pieces of the aircraft separated in flight and were not at the main impact site across the street from Tony and Vernell Schawang's farmhouse.

Specifically, the right wing separated roughly halfway along its length and the outboard section was found to the southeast and 2,503 feet (roughly half a mile) away. The vertical fin, with all of the left tailplane and part of the right (the two horizontal sections of the "T-tail"), was found east-southeast of the main site 4,375 feet, or approximately three-quarters of a mile, away. All major pieces of the aircraft were found by the teams of troops and other searchers within a one-square-mile area to the south and east of the main site.[1] At that main site, the major portion of the wreckage was "on a heading of approximately 110 degrees magnetic at impact, with the right wing low" just about facing back in the direction from which the jet came from Kansas City.

"Two funnel clouds were observed one-half mile southeast of the accident site approximately eight minutes after the accident" but there was no evidence of "hail damage, lightning strike or static discharge." Nor was there evidence of corrosion, fatigue or previous damage. All components were properly manufactured and formed to correct dimensions.² The cockpit and front door area were undamaged by fire, but there was extensive flame damage from the wheel well area back to the rear pressure bulkhead. The left wing was heavily fire damaged. The section of the right wing which separated and fell away showed no evidence of fire damage, but the stub-end left attached to the fuselage did, in varying degrees. The vertical fin separation occurred higher up the front side than the rear, and the rear side showed evidence of compression on the left. The portion of the fuselage frame which the front spar was attached to was pulled out of the fuselage frame and bent down counterclockwise and to the left. The fin landed upside down, with parts scattered away to the northwest. The tailplane trim actuator used to trim the aircraft's pitch was found in a position indicating it was set for a ¾ of a degree aircraft noseup position. Investigators and BAC's experts believed this trim setting was consistent with an airspeed of 260–280 knots at 5,000 feet mean sea level.³

Braniff's BAC One-11 operations manual gave specific instructions to pilots for operations in turbulence: "If severe turbulence cannot be avoided, the best airspeed, from all aspects of handling and strength, was 270 knots IAS (indicated air speed) up to 30,400 feet. Attitude flying is stressed and, in all cases, the autopilot should be engaged with the altitude hold switch "OFF." Pilots are also cautioned to remain at least five miles away from thunderstorms when operating below the freezing level." The trim setting was therefore consistent with Braniff's instructions regarding air speed in turbulence.⁴

"Examination of all control surfaces revealed no evidence of flutter or any significant pre-impact distress or malfunction." The same was true of the aircraft's hydraulic and electric generation systems. The two Rolls-Royce engines were still attached to the airframe; intense examination revealed nothing that would indicate any problems. Both engines' fire extinguisher bottles were still fully charged and were undeployed. There was no evidence that any of the cabin fire extinguishers were operated.⁵ In other words, everything on N1553 was operating as designed, with no malfunctions or flaws and within prescribed operating procedures. Whatever happened occurred almost instantaneously. If a fire had begun while passengers and crew were still conscious, the fire extinguishers would have been used or at least taken out of their regular storage positions, including the one in the cockpit. The board noted that "there is no evidence of any firefighting activities by the crew."⁶

Because the T-tail configuration was relatively new (the Aerospatiale Caravelle had a design which was closer to a "plus" than a "T," while the Douglas DC-9 and Boeing 727 were just coming into regular airline use), and because N1553's tail failed at altitude, the investigators paid special attention to the design and construction of this component: "It was found that the flight load requirements applicable to the BAC One–11 are basically the same as those that apply to all of the jet aircraft currently in the British and United States civil transport fleets." Those regulations required that "aircraft must be designed for certain flight maneuver and gust loads. The loads so specified are limit loads, the maximum loads expected in service. "However, the aircraft must also be capable of withstanding ultimate loads at least 50 percent greater than limit load before structural failure will occur."[7]

BAC designed its One–11 to withstand, below 20,000 feet, "limit derived gust velocities of 66 feet/second" for maximum gust intensity for 50 feet/second at design cruising speed and 25 feet/second at design dive speed.[8] While this is somewhat complicated terminology that tends to make everyone's eyes glaze over (unless you're an aeronautical engineer or jetliner nerd like this author), it gives a picture of the strength to which airliners in the U.S. and Britain must be built. The report noted that the term "derived gust velocity" did not mean "the actual or true velocity of a mass of moving air … [it] simply refers to an artificial gust of a specific shape which … will give accelerations generally in line with those which have been measured on similar aircraft in similar weather conditions." Such a theoretical basis for design "evolved over the years … [based on] the experience gained by monitoring past transport operation." As an example of that experience, the National Aeronautics and Space administration data for gust velocity data "was based upon nearly ten million nautical miles of experience."[9]

That data, according to the board, indicated "the design gust will be encountered once every 1,820 miles during actual flight in thunderstorms."[10] In other words, if a plane did nothing but fly around in thunderstorms *only*, then its design would be exceeded every 1,820 miles. But with the airlines' turbulence avoidance procedures, they wouldn't be flying in thunderstorms; they would successfully avoid them so that the design gust would not be encountered but once in 2.78 million nautical miles.[11]

Since it was known from almost the beginning that the aircraft broke up in the air, the board asked both British Aircraft and NASA's Langley Research Center to perform trajectory analyses on the parts which separated to calculate the path they took while falling through the air. The intersection of these paths will usually show at what point they separated from the aircraft.

The United States Air Force Air Defense Command radar had recorded Flight 250's track, ground speed and height above ground. BAC and NASA took the data and performed trajectory analyses (all in a way reminiscent of how Dr. Fujita worked) on both the separated parts and the larger portion of the aircraft that remained. From all this analysis, it was concluded that (1) "the breakup must have occurred within a very short time, perhaps in a time interval in the order of two seconds," and (2) "the fin-tailplane combination probably separated before the wing."[12] NASA and BAC also built a 1/40 scale model of the main body of N1553, dropped it from a simulated height of 6,400 feet, and recorded the data.

All of this work indicated that what was heard on the cockpit voice recorder was what they thought it was: N1553 broke up very quickly; the main fuselage entered a flat spin and took no more than 28 seconds to slam into Tony Schawang's soybeans. And the high-intensity gust force which broke off the tail and right wing was enough (based on the CVR tape, examination of bodies at the Prichard Auditorium, and the county attorney's determination that all 42 death certificates would list cardiac rupture as cause of death[13]) to almost instantaneously kill everyone aboard within two seconds.

Here again, early computer power was used to simulate various gust forces at different angles on the aircraft to see what it took to fail the tailplane and starboard wing. The lowest speed gust required to fail the tail was "calculated to be a 140 feet/second equivalent air speed (EAS) gust with a half-time of 0.125 seconds, applied from the right and angled upward 45 degrees." Other computer studies showed that once the tail was lost, the aircraft pitched downward so rapidly that the starboard wing failed. The investigators cautioned that the gusts derived from the studies were actual movements of air, not the design gust. They stated, "For example, the 140 feet/second EAS gust [derived from the computer modeling] would be equivalent to a combination of vertical and lateral derived gusts of 81.5 feet/second."

Taking into account the increase in airspeed heard on the CVR tape, the board also found that the failure of the tail would have occurred at 140 feet/second at 300 knots EAS," but it would have to be "nearly 158 feet/second" to fail the tail if the aircraft was at turbulence penetration speed. According to the wreckage investigation, it was being flown at that speed.[14]

The result of all these studies showed that the fin and tailplane failed almost simultaneously—and very, very suddenly. Those aboard Flight 250 probably experienced the increasingly bumpy ride, but almost surely never knew when they were knocked out of the sky.

The report then mentioned Dr. Fujita's involvement in weather studies and the conclusions in his testimony and three-part report.[15] The inclusion of his findings in the official report would indicate that the board accepted those findings as reasonable and likely.

The board ran through its findings and other research into turbulence. It concluded that the trip was conducted in a manner just like many other flights in similar conditions, but that airline personnel underrated the weather and did not pass any concerns to the crew. Captain Pauly was not advised of the intensity of the weather ahead, although he was told by the captain of another flight that radar wasn't showing a true picture of the intensity of the line of weather, which was as "long and mean as [he'd] seen in a long time."[16] Braniff dispatchers did not have this information, however. Weather Bulletin 447, found in the wreckage, forecast a few severe thunderstorms and numerous cumulonimbus clouds with tops up to 50,000 feet. While the dispatcher told the board he would not hesitate to ground a flight temporarily in accordance with company procedure, the weather on the night of August 6 did not warrant it. The board decided Flight 250 was dispatched "in good faith" but founded on an inaccurate weather situation analysis.

Once 250 was airborne and had already deviated from its original course "toward what appeared to be a hole in the line of clouds," First Officer Hilliker suggested deviating to Pawnee City to go around the squall line, but Captain Pauly decided to penetrate the hole shown on radar, to which he was already heading. The board felt that the captain's decision may have been different "had he known of the efforts of other crews to avoid penetrating this weather system." The dispatcher knew other flights which diverted or stayed on the ground and should have so informed Flight 250.[17]

The board ruled out a longitudinal upset as being the origin of the crash, as with Northwest 705 and the other incident in 1963-64. Braniff 250 didn't have time to be upset. It was hit by an extremely powerful force, probably the leading edge, horizontal rolling out in clear air ahead of the squall line. The board reported that relating the theoretical/design gust measurements to actual weather phenomena was "difficult since gusts of that severity have not been measured by man." The matter was "both one of definition and of effect on an aircraft." In a gust, "the aircraft will not have had time to adjust its flight path before it encounters the peak velocity and very large forces and accelerations can be produced. These, in turn, result in dynamic effects which raise the stresses in the structure to values which may be considerably higher than those which would result if the loads were applied slowly." While the precise velocities could not be computed beyond a general range, the board considered "that extreme turbulence was present and was, in fact, encountered by Flight 250" and that "it would have caused the failure of any modern civil transport."[18]

The report does give equal time to alternate investigations. Besides Dr. Fujita's findings, BAC, with the Airline Pilots Association (ALPA) put forth an alternate cause of the accident: a "complete loss of rudder feel which permitted the pilot to inadvertently apply full left rudder." There were 29 cases of sticking rudder feel system valves and they believed that reexamination of N1553's system would show that such a malfunction occurred. There were other points as well, so a joint FAA/industry team was put together. It could only find one documented case of a sticking rudder feel valve and no inflight loss of rudder feel. Board specialists examined ship 53's system again and could find no indication of failure. Other points of the BAC and ALPA dissent could not be replicated by the board.[19]

The dissent does bring up a curious, probably unanswerable, question: Why would BAC try to point away from a massive gust and toward failure of its own rudder feel system? And why would the pilots' union try to implicate the pilots of Flight 250? The first question may be that the public would think the One-11 could crash in turbulence, whereas a "rudder feel system" was unknown and could be "fixed" quickly and quietly. We can only speculate as to their thinking. As for ALPA, perhaps they sensed that Captain Pauly was about to get the full blame and they tried to deflect it onto something else. Regardless of this question, the board's conclusions were that the aircraft was strong, the gust was in the "extremely rare" category, and that, while it's possible he should have diverted west, Captain Pauly was proceeding using reasonable logic given what he knew or had been told (which wasn't everything). Neither he nor Hilliker nor the radar nor even the dispatcher could have seen a once-in-a-lifetime gust, capable of ending life within a second or two, coming towards them. At this point in its final report, the board made a few recommendations and hoped that computing power in the future would be able to both prevent accidents and help decipher the cause if they occurred, a wish that seems to have come mostly true. Finally, on pages 56–58, the board summed up its findings and made a statement of probable cause on page 59. It was, as usual, worded in carefully constructed ways: "The Board determines that the probable cause of this accident was inflight structural failure caused by extreme turbulence during operation of the aircraft in an area of avoidable hazardous weather."[20]

This statement can be broken down in a variety of ways, but the order in which it is constructed is simple. Flight 250 broke apart in extreme turbulence while being operated (by Braniff or the flight crew, however the reader wants to think of it) in avoidable hazardous weather. Given that probable cause and CAB/NTSB statements are not admissible in court and serve as fact-finding and advisory reports only, their language gets carefully parsed in almost every airliner accident case. Flight 250 was no exception. Still, there was now an explanation, a reasonably logical and defensible answer to the big *why* that had hovered over everyone's head for two years. It was time to move on, reluctantly and without forgetting but forward.

Postscript: Dawson

The Civil Aeronautics Board became the National Transportation Board on April 1, 1967, but the investigation of Flight 250 and its focus and personnel did not change. The new board was independent from the regulatory activities of the CAB.[21] A year later, the new NTSB released the final report on the crash of Braniff Flight 250 at Falls City. As mentioned, the basic idea behind such investigations and final reports is to (1) find out what happened, and (2) make recommendations to prevent it from happening again.

Just 15 days after the NTSB made its case for safer operations in turbulence, better information sharing between pilots and dispatchers and weather personnel, and the existence of clear air turbulence strong enough to rip apart brand new, state-of-the-art airliners, officials at Braniff International were filing the report away, relieved that the main ordeal was over. There would still be a few lawsuits to settle, but the boys in legal would handle all that; the suits would be settled or otherwise resolved and life would move on.

But at 4:48 p.m. on May 3, 1968, one of the airline's Lockheed L-188 Electras (called "Jet Power Electras" by Braniff) encountered an area of thunderstorms across its flight route between Houston and Dallas. The ride became worse and worse, and the decision, made possibly too late, to turn around and get out of Dodge as quickly as possible resulted in an unusual 180-degree retreat. But during the turn, made a mile from the small Texas city of Dawson, near Corsicana, the Electra disappeared in a midair explosion and fell to the ground in flames. Wreckage was found along a line stretching south-southeast up to north-northwest. An all-too-familiar scene began to be reenacted. Just as Bill Schock had done in Falls City, reporters in Dawson rushed out to the scene to take pictures and see things they would never forget. The death toll was high; 80 passengers and five crew members were all killed.[22]

The crash of Braniff International Flight 352, coming just 15 days after the final report on Braniff Flight 250 and under startlingly similar circumstances, was eerie, to say the least. After another lengthy investigation, a final report released June 19, 1969, declared that the probable cause of Flight 352's demise was "the stressing of the aircraft structure beyond its ultimate strength during an attempted recovery from an unusual attitude induced by turbulence associated with a thunderstorm. The operating in this turbulence resulted from a decision to penetrate an area of known severe weather." Therefore, Flight 352 possessed all the hallmarks of the accidents and incidents discussed here: an upset in severe turbulence and an attempted recovery which overstressed the aircraft, causing it to disintegrate in midair. On the surface, it would seem that no one, especially at Braniff, learned anything. The reality is more complex.

Investigators headed for Dawson and the process began all over again. Braniff changed some key operations and appeared to learn its lesson this time. Between the Dawson crash in May 1968 and its bankruptcy and shutdown in May 1982, the Flying Colors airline would not experience another fatal airplane crash, just the crash of the entire airline.[23]

Braniff International Airways Flight 250

Summary
Braniff Airways, Inc.
BAC One–11, N1553
7.6 miles north-northeast of Falls City, Nebraska
Saturday, August 6, 1966
Fatalities: 38 passengers, 4 crew; no survivors.

Report
National Transportation Safety Board
Aircraft Accident Report
SA. 393, File No. 1–0008
Adopted April 18, 1968

Fatalities

Crew
Captain Donald G. Pauly, 46, Minneapolis, MN
First Officer James A. Hilliker, 39, Bloomington, MN
Hostess Ginger Elaine Brisbane, 21, Minneapolis
Hostess Sharon Eileen Hendricks, 21, Minneapolis

Passengers
Bosted, Private Larry Joseph, Omaha
Broadfoot, Andrew Dewitt, Offutt AFB, Omaha
Chamblin, Nancy, Ft. Smith, AR
Chamblin, Susan, Ft. Smith, AR
Cox, Danny Ray, Omaha
Denies, Ronald L., Bayard, NE
Duerkson, Jean, Victoria, TX
Dyer, Ava, Washington, D.C.
Eschback, Donald, Omaha
Eskelinen, Kenneth, Omaha
Ferrero, Donald, Offutt AFB, Omaha
Foster, Leslie David, Jr., Omaha
Gilbertson, Patricia, North Little Rock, AR
Graeber, Lyman Monroe, Spring Park, MN
Gummers, Mrs. G., Omaha
Hamm, Mary Kay, Houston
Hamm, Susan, Houston
Howard, Charles E., Omaha
Hudson, Russell E., Ft. Worth
Jacobson, Patricia, Fargo, ND
Johnson, William O., Glen Flora, WI
Jordan, Cheryl Lyn, Minneapolis
Kowtaliw, Bohdan, Chicago
Kuhr, Mitchell L., Omaha
Kuhr, Ruth L., Omaha
Mayer, Adolph, Omaha
McConnell, Eugene P., Council Bluffs, IA
Mills, Opal, Gonzalez, TX
Murphy, William, Sauk Village, IL
Paul, John H., Overland Park, KS
Robertson, Garrett George, Omaha
Roettger, Grace Rhodes, Decatur, TX
Smith, Donald R., Bellevue, NE
Tejada, Virginia, Guatemala City, Guatemala
Ward, Charla J., Omaha
Welter, Robert D., Des Moines
Wilson, Frank, Fremont, NE
Wright, Donald Keith, Omaha

Part III. Postscript

20

New Normals: The "Before" and "After" Times

*It was not for closure. It was more for respect for my dad.
But closure? Very doubtful.*
—Bill Schock interview with Dan Hilliker

*And you danced away from the dinner table with him and we were all giggles
and it was so mushy. We were all so innocent. And I didn't think I'd be
happy till I found my way back there, to that moment—somehow.*
—Tim Hilliker, 2013

In Stillwater, Oklahoma, on the Thursday after the Flight 250 crash, a 22-year-old graduate student named Cheryl Lyn Jordan was laid to rest. She had joined Flight 250 at Tulsa and was traveling to Minneapolis, where she and her husband were making arrangements to move to the Twin Cities. They had met as undergraduates at the University of Tulsa and married in March 1965. They were moving north so that her husband, Jeff Jordan, could get to work. He recently had signed as a professional football player for the Minnesota Vikings.[1]

Sometime in the predawn darkness of August 7, 1966, a Braniff cargo agent from the Omaha Eppley Field station drove up to the chaotic scene on the Tony Schawang farm. As a Braniff employee with a badge, he probably was either not noticed or law enforcement on the site believed he was one of Braniff's official delegation. His name was Robert Kuhr, and his wife, Ruth, and five-year-old son, Mitchell, had taken a Braniff flight down to Kansas City earlier that morning so that they could visit family members, do some shopping, and spend time at the Kansas City Zoo. Their flights were on employee family passes arranged by Robert Kuhr, and he drove down from Omaha as soon as he heard about Flight 250's accident from dispatchers at Eppley Field.

Bill Schock would always remember some sights more than others from that terrible first few hours. One of them was seeing one of the hostesses lying on the ground in her Pucci mulberry-colored uniform, looking as if she were asleep, without a mark on her. And nearby was another unforgettable and poignant sight: a small boy, who had probably seated towards the front of the aircraft, lying in a similar manner to the hostess. He looked like he was asleep and there was no mark on him. Mitchell probably had experienced a busy day with the zoo and shopping and flying in his dad's airline's brand-new One–11.

Robert Kuhr found his wife and son in the debris and identified their bodies. They were taken out of the rain and mud to the Prichard Auditorium. The following Saturday, August 13, mother and son were given a double funeral at the First Methodist Church in Blair, Nebraska. They left behind husband and father Robert and an 18-month-old daughter and sister Melinda Sue, as well as Ruth's parents, Mr. and Mrs. Bernard Whipperman, of Arlington, Nebraska.[2]

Stories like these were multiplied 42 times. The Kuhr memorial was not the only double funeral. Mrs. G. Gummers was memorialized alongside her father, Adolph Mayer. There were two sets of sisters: Nancy Ann Chamblin, 18, and Susan Carol Chamblin, 15, both of Ft. Smith, Arkansas, who were on their way to Lincoln to visit an older sister; the other set of sisters were Mary Kay and Susan Hamm of Houston, Texas.[3]

In 2006, Daniel Hilliker of Eden Prairie, Minnesota, a suburb of Minneapolis, was a pharmaceutical systems consultant who was assigned a week of training in Omaha. Faced with a free weekend between sessions, he made a decision. He e-mailed the *Falls City Journal* and arranged to meet the legendary Bill Schock for help with getting more details on the crash and actually visiting the soybean field. He rented a car at Omaha Eppley Field and drove the 90 miles south.[4]

As with my own visit this last October, Bill took Dan in as a long-lost friend. He showed Dan everything on the crash he could get hold of, the same things he would show me later. And since he is Bill Schock, he immediately whipped out his reporter's notebook and camera and interviewed for the front page of the *Journal*. He did the same thing with me during my visit.

Dan Hilliker has lived with the very public loss of his father since he was 15 years old. On the same summer Sunday morning when his little brother Tim was hopping down the stairs to watch *Gene Autry* on television, Dan was up in his bedroom asleep. Sometime in the predawn darkness, their mother, Patricia, answered the front door in a fog of sleep. She thought it might be her husband, Jim, home already from his run back from New Orleans, but later she said she knew something was not right.

When she opened the front door, she nearly collapsed. There stood three impeccably dressed Braniff pilots sporting the new black uniforms which signified Harding Lawrence's new broom. The three were from the Minneapolis station, and they obviously had come with bad news. Jim was supposed to be home by now; after a short sleep he would make breakfast for early birds in the neighborhood. Milkmen, newspaper boys, anyone who wanted to drop by would get a welcome and some doughnuts or waffles or whatever was on. Reality was growing on Pat; there would be no more breakfasts. She collapsed into a dining room chair, holding it together long enough to call her brother, Bob Lindseth, for help. He came over as quickly as he could drive the distance.

Pat told her brother after he arrived that he would have to tell the four kids; she couldn't get the words out. Bob headed upstairs. He would start with Dan, the oldest. Dan's reaction was sleepy, shocked disbelief. He yelled at Bob, "I don't believe you!" He thought it was a nightmare (it was) and may have even taken a swing at his uncle—nothing serious, just a grieved reaction. Bob told Kathleen, 12, next; she ran downstairs and out behind the garage and was sick.[5]

When nine-year-old Tim Hilliker came down a bit later expecting his Gene Autry Sunday morning, his Uncle Bob instead pulled him away from the living room into a downstairs bathroom, knelt down and told him his father would not be coming home. Tim remembers all these years later looking at his mother, sitting in a dining room chair and staring into the living room. It was the "new normal," the "after" time, defined by the "normal before" the crash and the "new normal after" the crash. Everything since that Sunday morning of August 7, 1966, has been "after."[6]

Once wreckage removal was complete and all the strangers had left, quiet and peace and normality returned to Falls City. But for Tony and Vernell Schawang, it was anything but normal. They were so traumatized by the events of August 1966, they experienced trouble sleeping and even staying in the house. Everywhere they turned was a reminder; outside across the street a somewhat charred, barren spot in the soybeans stared back at them. They stayed with others at first and then sold the farm completely and moved over near Shubert so they wouldn't be continually seeing a fireball and wreckage and bodies and their home used as a command post. Tony died in 1988; Vernell passed away in 2005.

The farm was in other hands for a number of years before being brought back into the family. Tony's nephew, Kenny Schawang, and his wife, Elaine, bought the place in the 1980s, renovated the house and live there now. The day after I met Bill Schock at the *Falls City Journal* in October of 2012, Kenny and Elaine graciously met me at the farm after work.

They both were among the 300 witnesses of the crash that wild August night; they were not married yet and were still teenagers. Kenny was at his home several miles away getting ready for bed; Elaine was babysitting some neighbors' children. Like many others, she was standing on the neighbors' front porch and looking towards the storm, so she experienced the fireball and the aftermath.[7] Their grandchildren now play in the kitchen where Kenny's Aunt Vernell brewed so much coffee for so many people and where his Uncle Tony sat and spoke to Claire Hurlbert and Bill Schock while chaos swarmed around them in the early morning darkness and the lightning flashed overhead.

Kenny and a neighbor constructed and welded a permanent marker to stand as a memorial. It was placed at the corner of the two intersecting roads at the edge of the main crash site, and the inscription reads:

> In memory of the 4 crew members and 38 passengers of Braniff Flight #250 who perished in this field on August 6th, 1966.
> John 14:1–3
> On behalf of the residents of Richardson County.[8]

The scripture reference is to a passage in the New Testament in which Christ encourages his followers: "Do not let your hearts be troubled. You believe in God; believe also in me. My Father's house has many rooms; if that were not so, would I have told you that I am going there to prepare a place for you? And if I go and prepare a place for you, I will come back and take you to be with me that you also may be where I am."[9]

Elaine Schawang takes care of the marker, pulling weeds and freshening wild flowers and making sure things are in apple pie order. At the base of the memorial is an embedded piece of N1553 not needed in the investigation, a jagged piece of metal that looks like it was an actuator rod of some sort. Over the years, such pieces have occasionally been pulled up out of the fields along the entire breakup path.

In 2006, at a time which coincidentally came just before the 40th anniversary of the crash of Braniff 250, Dan Hilliker continued his tour of Falls City, thanks to Bill Schock, and saw all the "sights." He saw the Prichard Auditorium where his father's body was taken with all the others, and the offices of the *Falls City Journal*, the Hotel Stephenson, home to investigators and family members, and so on. Dan also saw the large shelter pavilion in City Park. In appreciation for what the Falls City community did in the summer of '66, Braniff Airways donated a check for $5,000. The city used it to build the shelter pavilion, which has shielded a million picnics and reunions and other events.

Finally, on a Saturday morning, Bill and Dan drove out to the farm where Tony and Vernell Schawang watched their world come apart. Kenny joined them at the site, telling Dan what he knew and experienced and what stories were told at the time. As Bill Schock noted, "Occasionally it got to be too much for the copilot's son. He moved away." But he wanted to hear every little detail. "He walked down to the site, about 50 yards from the east-west road (Road 714) and remained there for a time. When he returned he was provided with a piece of the wreckage that had been picked up years ago. Then we returned to town, and he started on his trip back to Omaha."[10]

On the night of August 6, 1966, Bill Schock and a few others saw Jim Hilliker and Don Pauly in the shattered lower remains of the cockpit. Bill said he would never forget that Jim's left hand was still gripping the control column; he and Captain Pauly were still flying the aircraft when the sudden blow fell and froze them. And on Jim's left hand gripping the yoke was his wedding ring, clearly visible in the lights being played over the wreckage.[11]

Immediately after the crash, Jim's father, Spike Hilliker, and Jim's brother, Bob Hilliker (a different Bob, his brother-in-law Bob Lindseth, stayed in Bloomington with Pat and the kids), drove down to Falls City. They identified Jim and were given the wedding ring so it could be returned to Pat. Pat kept it for many years, until their oldest, Dan, got married. During his wedding to his wife Sandy, Jim Hilliker's ring was placed on Dan's finger. It survived the crash of Flight 250 and it was now a visual, tangible connection between Dan and his father.[12]

Now it was years later and Dan was ready to leave Falls City after his first visit. He first stopped at a florist shop and then took a bouquet of flowers back out to the Schawang farm. He sat down for an hour and took in the quiet, pastoral scene. He spoke with Kenny and Elaine's son, Korey, asking him to give the flowers to Korey's mother, and then he left for Omaha. "In departing, he said, emotionally that it had been a very memorable day for him. Forty years certainly hadn't put the tragedy out of mind. But he had seen what he had wanted to see and read what he had wanted to read. And he had talked with people who had been there," reported Bill Schock.

Prior to his visit, Dan had called his wife, Sandy, to tell her about his Falls City plans. "She told him, sympathetically, that perhaps it would mean closure for him. "'It was not for closure ... it was more for respect for my dad.' "But closure?" asked Dan Hilliker. "Very doubtful."[13]

Shortly after Dan's visit in the summer of 2006, Falls City came together to remember Braniff Flight 250. Just as in 1966, so many people organized and worked hard to make the 40th anniversary a respectful (and this time less stressful) occasion. Several family members of Flight 250's passengers and crew attended, including Dan Hilliker and his siblings, Kathy, Tim and David. They found a warm embrace from the community. They saw photos they didn't know existed and heard stories they didn't know were told, and they met people who treated their loved ones with dignity and respect. There was also an interfaith church memorial service and lunch; tickets sold out. And there was a visit out to the farm. The occasion was likely the most people at the site at one time since the crash itself.[14]

In 2010 a chance encounter via the Internet led me to connect to Tim Hilliker, Jim and Pat's third child and the one who bounded down the stairs on that Sunday morning only to find that the "after" time was already underway. I also was graciously given a few hours one afternoon in Eden Prairie, Minnesota, to meet Dan and Sandy and sister Kathy and Uncle Bob. (Pat Hilliker passed away in 2003.)

I had enjoyed numerous phone calls with Tim, and now I sat around Dan and Sandy's table with the friendliest golden retriever in the world and delicious chocolate chip cookies as Dan and Kathy and Bob described Jim Hilliker, showing me pictures and his pilot's wings (the one he was wearing on Flight 250 has a corner bent permanently backwards). They remembered all the great times with their father and how he loved to make jokes and tease people and have fun, especially in the cockpit. Jim had told his children about one of his favorite cockpit jokes;

Jim Hilliker, probably in the 1950s just after his employer, Mid-Continent Airlines, was absorbed by Braniff International (courtesy Hilliker family).

variations on it were, at least back in the golden age of airline operations, used by many a veteran flight crew on unsuspecting flight attendants. Jim and the captain of a particular flight discovered they would have a new hostess in her first few weeks of flying on one of their segments. She was working the forward part of the cabin, while the more experienced and senior hostess worked the rear. Jim told the new hire after takeoff that they had encountered problems with the flush mechanism of the forward lavatory, but that it could be flushed manually via a hand crank set in the floor of the cockpit. This crank was actually used to manually lower landing gear in case of hydraulic failure; if it wasn't active, the crank could be spun, but nothing would happen and no harm would be done. Jim instructed the new hostess that every time someone went to the lavatory, she was to enter the cockpit and spin the crank several times to flush the toilet.

The new hostess proceeded to run off her feet. She would serve a drink, then run into the cockpit and spin the hand crank around and around, then run back into the cabin to serve more drinks. By the time they landed at their destination, she was completely exhausted, but the flight crew told her she did an excellent job. The senior hostess told the new one that she did a fine job with the service, but wondered why she kept running into the cockpit so much. It was then that she heard Jim Hilliker and the unnamed captain laughing loudly in the cockpit. The senior hostess set her new colleague right and then visited a terrible, but unnamed, vengeance on Jim and the captain.[15]

Jim never met a stranger, was protective and loyal to his crew, and laughed and loved life. His story was one of 42 which stopped at 11:12 p.m. on August 6, 1966.

In the backyard of Dan Hilliker's home in Eden Prairie is a small ornamental fountain which belonged to their mother. Hanging from a tree nearby is the piece of N1553 recovered from the Nebraska field by Kenny Schawang and given to Dan in 2006. Dan painted it bright yellow and it swings in the breeze. Every day now when Dan goes into his backyard, he pats the fountain statue on the head and says, "Good morning, mom!" and then he reaches up and touches the part of his father's BAC One–11 and says, "Good morning, dad!"[16]

Jim and Pat Hilliker now have several grandchildren; great-grandchildren, as these things go, are probably fairly close at hand. And life goes on.

Notes

Preface

1. Bijan Vaigh, Ken Fleming and Thomas Tacker, *Introduction to Air Transport Economics: From Theory to Applications*, 2nd ed. (Burlington, VT: Ashgate, 2013), quoted in chapter 13.

2. Health Information Resource Database, "Home Safety Council: Safe Kids Worldwide," http://www.health.gov/NHIC/NHICScripts/Entry.cfm?HRCode=HR3768 (accessed April 1, 2013).

Chapter 1

1. *Elyria* (OH) *Chronicle-Telegram*, February 4, 1959, p. 2.
2. Ibid.
3. Ibid., 1.
4. Ibid.
5. Ibid.
6. Larry Lehmer, *The Day the Music Died* (New York: Shirmer, 1997).
7. *Cedar Rapids* (IA) *Gazette*, February 4, 1959, p. 36.
8. Civil Aeronautics Board (CAB), Report on Pan American Flight 115, Report No. US COMM-DC-25158/1–0006 (hereafter CAB PA115), November 3, 1959, p. 10.
9. *Elyria Chronicle-Telegram*, 2.
10. CAB PA115, 10.
11. Ibid., 1.
12. Ibid.
13. *Cedar Rapids Gazette*, 36.
14. Ibid.
15. *Middletown* (NY) *Daily Record,* in Leonard Lyons, "The Lyons Den," February 13, 1959, p. 35.
16. *Cedar Rapids Gazette*, 36.
17. CAB PA115, 2.
18. Ibid.
19. Ibid.
20. Ibid., 3.
21. Ibid., 2.
22. Ibid.
23. Ibid., 3.
24. Ibid.
25. Ibid.
26. Ibid.
27. Ibid.
28. Ibid.
29. *Cedar Rapids Gazette*, 37.
30. CAB PA115, 3.
31. Ibid., 4.
32. Ibid.
33. Ibid.
34. Ibid.
35. Ibid., 9.
36. Ibid., 1.
37. Clive Hirschhorn, *Gene Kelly: A Biography* (New York: St. Martin's, 1985).
38. Harry Kurnitz, *Reclining Figure* (New York: Dramatist's Play Service, 1998).
39. Susan Oliver, *Odyssey: A Daring Transatlantic Journey* (New York: Macmillan, 1983).
40. Ibid., 73.

Chapter 2

1. "Boeing 707-'Year One'-1960-Part I-V," YouTube video, posted February 19, 2009, http://www.youtube.com/watch?v=ph1FXcIqnNg (accessed February 15, 2013).
2. Ibid.
3. Rod Steiner and Martin R. Copp, "A Review of Atmospheric Turbulence and Its Sig-

nificance to Jet Transport Operations" (NASA, 1958), 119.
4. Ibid.
5. Ibid.
6. Ibid., 122.
7. Ibid., 123.
8. Northwest Orient Airlines, System Timetable, February, 1963, p. 1.
9. *Old Farmer's Almanac,* Miami and Chicago weather, February 12, 1963 (accessed February 3, 2013), http://www.almanac.com/weather/history/FL/Miami/1963-02-12.
10. Northwest, timetable, 6.
11. Civil Aeronautics Board, June 4, 1965, Report on Northwest Airlines Flight 705, Report No. SA372/1-0006 (hereafter CAB NW705), 3.
12. *Mt. Vernon* (IL) *Register-News*, February 13, 1963, p. 1.
13. Northwest, timetable, 6.
14. CAB NW705, ii.
15. Ibid., 7.
16. Ibid.
17. Ibid., ii.
18. Ibid., 1.
19. Ibid., i.
20. Ibid., 19.
21. Ibid., i.
22. Ibid.
23. Ibid.
24. Ibid., 4.
25. Ibid.
26. Ibid., 3.
27. Ibid., 4.
28. Ibid., 3.
29. Ibid., 4.
30. Ibid.
31. Ibid., 2.
32. Frank A. Tinker, *Popular Mechanics*, June 1963, p. 103.
33. CAB NW705, p. 2.
34. Ibid.
35. Ibid.
36. Ibid.
37. Ibid.

Chapter 3

1. Civil Aeronautics Board, June 4, 1965, Report on Northwest Airlines Flight 705, Report No. SA372/1-0006 (hereafter CAB NW705), 3
2. Ibid.
3. Ibid., 2.
4. *Mt. Vernon* (IL) *Register-News*, February 13, 1963, p. 1.
5. Ibid.
6. CAB NW705, iii.
7. *Mt. Vernon Register-News*, 1.
8. CAB NW705, 5.
9. *Mt. Vernon Register-News*, 1.
10. CAB NW705, 5.
11. Jim Bishop, "Jim Bishop: Reporter," *San Antonio Light*, May 17, 1963, p. 53.
12. CAB NW705, p. 5.
13. Ibid., 6.
14. Ibid.
15. Ibid.
16. Ibid.
17. CAB NW705, pp. 4–5.
18. Ibid.
19. CAB NW705, p. 5.
20. CAB NW705, p. 7.
21. Ibid.
22. Ibid.
23. Ibid.
24. CAB NW705, pp. 7–8.
25. Ibid., 9.
26. Ibid., 14.
27. Ibid.
28. CAB NW705, p. 15.
29. Ibid., 19.
30. Ibid.
31. Ibid.
32. Ibid.
33. Ibid.
34. Ibid., 20.
35. Ibid., 21.
36. Ibid.
37. Ibid., 22.

Chapter 4

1. Warren R. Young, *Life*, December 18, 1964, 86.
2. International Civil Aviation Organization (ICAO), Reports on Air Transportation in 1963, press release, Montreal, Canada, May 1, 1964.
3. Young, *Life*, 86.
4. Ibid., 90.
5. Ibid.
6. Ibid.
7. ICAO, press release.
8. Young, 91.
9. Ibid., 90.

Chapter 5

1. *Bakersfield Californian*, July 13, 1963, p. 26.

2. *Roselle* (IL) *Register*, November 9, 1951, p. 1.
3. *Bakersfield Californian*, July 13, 1963, p. 1.
4. United Air Lines, timetable, Summer 1963.
5. United Air Lines, advertisement, "The Chicago Executive," 1955.
6. *Bakersfield Californian*, July 13, 1963, p. 1.
7. *Pasadena Independent Star-News*, July 14, 1963, p. 3B.
8. *Bakersfield Californian*, 1.
9. S. Ragland, Jr., R.M. Chambers, R.J. Crosbie, L. Hitchcock, Jr., "Simulation and Effects of Severe Turbulence on Jet Airline Pilots," Johnsville, PA, August 13, 1964, ii.
10. Ibid.
11. Ibid.
12. Ibid.
13. Ibid.
14. Ibid., iii.
15. Ibid.

Chapter 6

1. Eastern Air Lines, System Timetable, July 1, 1963, p. 1.
2. Jack E. Robinson, *Free Fall: The Needless Destruction of Eastern Air Lines and the Valiant Struggle to Save It* (New York: HarperBusiness, 1992).
3. Eastern Air Lines, timetable, 60.
4. *Flight*, "DC-8," November 18, 1960, p. 803.
5. Boeing Company, "DC-8 History," http://www.boeing.com/history/mdc/dc-8.htm (accessed February 15, 2013).
6. Ibid.
7. Douglas Aircraft Company, "Breaking the Sound Barrier," press dispatch, Wednesday, August 21, 1961, p. 1.
8. "Douglas DC8 – 'Birth of a Jet'-1959- Part I-II," YouTube video, February 19, 2009, http://www.youtube.com/watch?v=U_oC-ijfQkA (accessed February 13, 2013).
9. *Flying*, "Test Piloting the DC-8, June, 1959," p. 2.
10. Ibid.
11. Ibid., 3.
12. PPG, "Facts About PPG," Brochure, 2012, p. 1.
13. National Aviation Hall of Fame, "Harold Pitcairn Biography," http://www.nationalaviation.org/pitcairn-harold/ (accessed February 12, 2013).
14. Ibid.

15. *San Antonio Light*, May 15, 1977, p. 9.
16. Eastern, timetable, February 1, 1964, p. 2.
17. Calculations, Measuring Worth, http://www.measuringworth.com/ (accessed February 23, 2013).

Chapter 7

1. Fred McClement, *It Doesn't Matter Where You Sit* (New York: Holt, Rinehart, Winston), 143; Carol Miller, *Life*, December 6, 1963, p. 23.
2. *Old Farmer's Almanac*, Jamaica, NY, weather, November 9, 1963, http://www.almanac.com/weather/history/NY/Jamaica/1963-11-09 (accessed February 15, 2013).
3. Fred Cox DC-8 Jet Collection, Eastern Air Lines Fleet Information, http://www.dc-8jet.com/eal-dc8-fleet-info1.htm (accessed February 14, 2013).
4. McClement, 143.
5. Miller, *Life*, December 6, 1963, p. 23.
6. Ibid.
7. Ibid.
8. Ibid.
9. McClement, 144.
10. Miller, *Life*, p. 23.
11. *San Antonio Express*, November 12, 1963, p. 1.
12. Ibid.
13. *Hutchinson* (KS) *News*, November 21, 1963, p. 28.
14. Miller, *Life*, 23.
15. *Pampa* (TX) *Daily News*, November 11, 1963, p. 1.
16. Miller, *Life*, 23.
17. Ibid.
18. Ibid.
19. *Pampa Daily News*, 1.
20. Ibid.
21. Civil Aeronautics Board, June 4, 1965, Report on Eastern Air Lines Flight 304, Report No. 1 SA379/1-0006, p. 3.
22. *Hutchinson News*, 28.
23. Ibid.
24. Young, *Life*, December 18, 1964, p. 87.
25. *Pasadena Independent*, November 13, 1963, p. 2.

Chapter 8

1. Civil Aeronautics Board, June 4, 1965, Report on Eastern Air Lines Flight 304, Report

No. SA379/1-0006 (hereafter CAB EA304), p. 1.
2. Ibid., 3.
3. Ibid.
4. Ibid.
5. Eastern, timetable, February 1, 1964, pp. 31–32.
6. CAB EA304, pp. 2–3.
7. Ibid., 1–2.
8. Ibid., 5–6.
9. Ibid., 5.
10. Ibid., 6.
11. Ibid., 2.
12. Ibid., 7.
13. Ibid.
14. Ibid., 2.
15. Ibid.
16. Ibid.

Chapter 9

1. Civil Aeronautics Board, June 4, 1965, Report on Eastern Air Lines Flight 304, Report No. SA379/1-0006 (hereafter CAB EA304), 7–8.
2. Ibid.
3. *Billings Gazette*, February 26, 1964, p. 1.
4. Ibid., 8.
5. Arthur Hailey, *Airport* (New York: Doubleday, 1968).
6. *Madison Wisconsin State Journal*, February 26, 1964, p. 2, section 1.
7. CAB EA304, p. 7.
8. *Joplin Globe*, February 26, 1964, p. 1A.
9. *Pasadena Independent*, February 26, 1964, p. 3.
10. Ibid.
11. *Kenneth Lee Spencer*, Internet Movie Database, http://www.imdb.com/name/nm0818007/bio (accessed February 4, 2013).
12. *French Senate Almanac Online*, "Marie-Hélène Lefaucheux," http://www.senat.fr/senateur-4eme-republique/lefaucheux_marie_helene0541r4.html (accessed February 2, 2013).
13. *Madison* (WI) *Capital Times*, October 22, 1941, p. 10.
14. Ibid.
15. *Madison Capital Times*, May 10, 1942, p. 12.
16. Platteville State Teachers College, *The Pioneer*, 1941 yearbook.
17. Ibid.
18. *Wisconsin State Journal*, February 15, 1943, 14.
19. Ibid.
20. Ibid., April 8, 1943, p. 11.
21. Ibid., November 3, 1943, p. 10.
22. Ibid., December 8, 1943, p. 11.
23. 351st Bomb Group Historical Society, http://351st.org (accessed February 2, 2013).
24. Ibid.
25. Ibid., Crew Member Search Page, Grant Newby, SN 0-815211.
26. Ibid.
27. Ibid.
28. *Wisconsin State Journal*, July 6, 1944, p. 15.
29. 351st Bomb Group Historical Society.
30. Ibid.
31. Ibid.
32. *Wisconsin State Journal*, July 16, 1944, p. 17.
33. Ibid., March 9, 1945, p. 7.
34. Ibid., May 27, 1946, p. 12.
35. Colloquial, origin unknown.
36. CAB EA304, p. 3.
37. *French Senate Almanac Online*, "Marie-Hélène Lefaucheux."
38. Ibid.
39. Ibid.
40. *Chicago Star*, December 11, 1951, p. 17.
41. *Abilene Reporter News*, January 12, 1953, p. 4A.
42. *Winnipeg* (Manitoba) *Free Press*, June 22, 1957, p. 3.
43. *Joplin Globe*, February 26, 1964, p. 1A.
44. *Altoona Mirror*, March 19, 1964, p. 22.
45. Larry Collins and Dominique Lapierre, *Is Paris Burning?* (New York: Pocket Books, 1965).
46. Bosley Crowther, "The Screen: Is Paris Burning? Takes Great Documentary Material and Turns It into a Garble," *New York Times*, November 11, 1966, p. B1.
47. *French Senate Almanac Online*, "Marie-Hélène Lefaucheux."
48. *Fayette* (IA) *County Leader*, March 23, 1939, p. 4.
49. Ibid.
50. Ibid.
51. *Fayette County Leader*, April 11, 1940, p. 1.
52. Brian Locke, "Strange Fruit: White, Black, and Asian in the WWII Combat Film Bataan," *Journal of Popular Film and Television* 36, no. 1 (Spring 2008), 9.
53. *Cabin in the Sky*, Internet Movie Database, http://www.imdb.com/title/tt0035703/ (accessed January 25, 2013).
54. Locke, "Strange Fruit," 10.
55. Ibid.
56. Ibid., 11.
57. Ibid., 14.

58. Ibid., 15.
59. National Lawyers Guild, *Civil Liberties Docket* 8, no. 1 (November 1962).
60. *Altoona Mirror*, March 19, 1964, p. 22.

Chapter 10

1. Civil Aeronautics Board, June 4, 1965, Report on Eastern Air Lines Flight 304, Report No. SA379/1-0006 (hereafter CAB EA304), 7.
2. Ibid.
3. Ibid.
4. Ibid.
5. Ibid.
6. *Titusville Herald*, February 27, 1964, p. 1.
7. CAB EA304, p. 7.
8. Ibid.
9. Ibid., 25.
10. Ibid., 5.
11. National Air Transportation Association, "Aircraft Maintenance and System Technology Committee Best Practices: Minimum Equipment List (MEL)," Alexandria, VA, Fall 2012, p. 1.
12. H.H. Hurt, Jr., *Aerodynamics for Naval Aviators* (1943), 250.
13. Ibid.
14. Ibid.
15. Ibid.
16. CAB EA304, p. 10.
17. Ibid., 4.
18. Ibid.
19. Ibid.
20. Ibid.
21. CAB EA304, pp. 4-5.
22. Ibid., 5.
23. Ibid.
24. National Transportation Safety Board, Report on Alaska Airlines Flight 261, December 30, 2002, p. 1.
25. CAB EA304, pp. 4-5.
26. Ibid., 5.
27. Ibid.
28. Ibid., 23.
29. Ibid., 16-17.
30. Ibid., 22.
31. Ibid.
32. S. Ragland, Jr., R.M. Chambers, R.J. Crosbie, L. Hitchcock, Jr., "Simulation and Effects of Severe Turbulence on Jet Airline Pilots," NADC-ML-64 11, Johnsville, PA, August 13, 1964, ii.
33. CAB EA304, p. 23.
34. R.A. Hess, "Unified Theory for Aircraft Handling Qualities and Adverse Aircraft-pilot Coupling," 1997, p. 1141.
35. CAB EA304, pp. 23-24.
36. Ibid., 22.
37. Ibid., 24.
38. Ibid.
39. Ibid.
40. Ibid.
41. CAB EA304, p. 26.
42. CAB EA304, p. 27.
43. *Altoona Mirror*, March 19, 1964, p. 22.
44. Robinson, *Free Fall*, 1.
45. Eastern Air Lines Flight 304 Memorial Cenotaph, http://www.findagrave.com/cgi-bin/fg.cgi?page=gr&GRid=44981629 (accessed February 12, 2013).

Chapter 11

1. Anthony Sampson, *Empires of the Sky* (New York: Random House, 1985), 10.
2. Ibid.
3. John J. Nance, *Splash of Colors: The Self-Destruction of Braniff International* (New York: William Morrow, 1984), 1-5.
4. Geza Szurovy, *Classic American Airlines* (Minneapolis: MBI, 2003), 159-160.
5. Ibid.
6. Sampson, *Empires of the Sky*, 25.
7. Szurovy, *Classic American Airlines*, 163.
8. Nance, *Splash of Colors*, 25.
9. Ibid., 35.
10. Braniff International Airways, Inc., Annual Company Report, 1965, p. 2.
11. Ibid., 3.
12. Nance, *Splash of Colors*, 40.
13. "Kubrick2001: The Space Odyssey Explained," http://www.kubrick2001.com/ (accessed March 15, 2013).
14. Robert Gandt, *Sky Gods*, 2.
15. GlobalSecurity.org, "Boeing 2707 SST Competition," http://www.globalsecurity.org/military/systems/aircraft/b2707-comp.htm (accessed March 18, 2013).
16. Hiller Aviation Museum, "Exhibits," http://www.hiller.org/exhibits.shtml (accessed March 18, 2013).
17. *Sports Encyclopedia Online*, "Seattle Supersonics (1967-2008), Nickname," http://www.sportsecyclopedia.com/nba/seattle/sonics.html (accessed March 3, 2013).
18. Nance, *Splash of Colors*, 115.

Chapter 12

1. Mary Wells Lawrence, *A Big Life in Advertising* (New York: Simon & Schuster, 2003), 34.
2. Ibid.
3. Ibid.

4. Ibid., 36.
5. Ibid.
6. "Braniff Airways 'End of the Plain Plane,'" YouTube video, posted November 10, 2006, http://www.youtube.com/watch?v=H3_aNtQFsLk (accessed February 25, 2013).
7. *Waterloo Daily Courier*, full-page advertisement, "The End of the Plain Plane," November 30, 1965, p. 7.
8. "Colors Are Fun," *Time*, April 15, 1966, p. 90.
9. "Braniff International Air Strip," Flight.org video and advertisements, posted by Marty, March 9, 2010, http://www.flight.org/blog/2010/03/25/braniff-international-air-strip/ (accessed March 2, 2013).
10. Ibid.
11. *Time*, "Colors Are Fun," 90.
12. Ibid.
13. Ibid.

Chapter 13

1. "British Aircraft Corporation One-Eleven," *Aircraft Engineering Journal* (September 1964), 2.
2. Braniff International Airways, Inc., Annual Company Report, 1965, p. 4.
3. Ibid.
4. Ibid.
5. Massachusetts Institute of Technology, "Airline Data Project," Global Airline Industry Program, http://web.mit.edu/airlinedata/www/default.html (accessed March 25, 2013).
6. Ibid., Glossary, http://web.mit.edu/airlinedata/www/Res_Glossary.html (accessed March 25, 2013).
7. Ibid.
8. Ibid.
9. Ibid.
10. Braniff International Airways, Inc., Annual Company Report, 1970, p. 3.
11. "British Aircraft Corporation One-Eleven," *Aircraft Engineering Journal* 34, no. 5 (September 1964), 2.
12. Ibid., 1.
13. British Aircraft Corporation, BAC One-Eleven sales brochure, 1963, p. 2.
14. Ibid.
15. *Aircraft Engineering Journal*, 1.
16. British Aircraft Corporation, BAC One-Eleven sales brochure, 1963, p. 4.
17. Timothy Walker, *The First Jet Airliner: The Story of the De Havilland Comet (Aircraft of Distinction)* (Newcastle upon Tyne: SCOVAL, 2007), ii.
18. British Aircraft Corporation, BAC One-Eleven sales brochure, 1963, 5.

19. Ibid.
20. *Aircraft Engineering Journal*, 19.
21. Ibid.
22. Brian Trubshaw, *Concorde: The Inside Story* (London: Sutton, 2003), xvii.
23. *Aircraft Engineering Journal*, 2.
24. Ibid.
25. Braniff International Airways, timetable, July 1966, p. 15.
26. "Air Transport: Braniff One-Eleven Plans," *Flight International*, April 22, 1965, p. 632.
27. "Air Transport: Braniff One-Eleven Decor," *Flight International*, April 22, 1965, p. 639.
28. Timetables
29. *Aircraft Engineering Journal*, September 1964, 2.

Chapter 14

1. Martin Bowman, *B-17 Groups of the 8th Air Force in Focus* (Walton-on-Thames: Red Kite, 2004), 58.

Chapter 15

1. Civil Aeronautics Board, April 18, 1968, Report on Braniff International Airways Flight 250, Report No. SA393–1-0008 (hereafter CAB BI250), Appendix B, 1–2.
2. See Jon Proctor's photo of N1553 in this volume.
3. CAB BI250, Appendix B, 1–2.
4. Ibid.
5. CAB BI250, Appendix A, 1–2.
6. CAB BI250, 4.
7. See chapter 12.
8. See chapter 13.
9. *Huron* (SD) *Daily Plainsman*, "Many of 42 Victims Were on Visits," August 8, 1966, p. 1.
10. CAB BI250, 5.
11. Ibid.
12. Ibid., 6–7.
13. Ibid., 5.
14. *Burlington* (IA) *Hawk-Eye*, "Mother, Son Among Dead," August 8, 1966, p. 2.
15. CAB BI250, 3–4.
16. Ibid.
17. Ibid.
18. Ibid.

Chapter 16

1. *Falls City Journal*, "Experience a Terrible Nightmare," August 8, 1966, p. 1.
2. Ibid.

3. *Falls City Journal*, "42 Dead in Airliner Crash; Crash Scene Ghastly Sight," August 8, 1966, p. 1.
4. Ibid., 2.
5. Steve Pollock interview with Bill Schock, Falls City, Nebraska, October 26, 2012.
6. National Transportation Safety Board, e-mail to Steve Pollock, "All of this public docket [Files related to the investigation of Braniff Flight 250] was destroyed," July 25, 2012.
7. *Falls City Journal*, "42 Dead," 2.
8. *New York Times*, "All 42 Die as Jet Falls in Flames in Nebraska Field; Witnesses Tell of Flash and Explosion Before Braniff Airliner Hit the Ground; Plane Built in Britain; U.S. Investigators to Study Wreck of First BAC 111 to Have Accident in Nation," August 8, 1966, p. 1.
9. *Falls City Journal*, "Big Roles by Civil Defense, Guards," August 8, 1966, p. 1.
10. *Chicago Tribune*, "2 Recorders Hold Clews to Air Crash; Probe Debris and 42 Deaths," August 8, 1966, p. 1.
11. *Falls City Journal*, "42 Dead," p. 2.
12. *Falls City Journal*, "Big Roles by Civil Defense, Guards," August 8, 1966, p. 1.
13. *Falls City Journal*, "42 Dead," p. 2.
14. *Falls City Journal*, "Air Traffic Congested Over Site of Tragedy," August 8, 1966, p. 2.
15. *Falls City Journal*, "42 Dead," p. 2.
16. *Falls City Journal*, "Big Roles by Civil Defense, Guards," August 8, 1966, p. 1.
17. *Falls City Journal*, "Many Persons Help in County Tragedy," August 9, 1966, p. 2.
18. J. Edgar Hoover, Director, Federal Bureau of Investigation, United States Department of Justice, Letter of September 26, 1966, to Walter Henshel, Vice President of Public Relations, Braniff International Airways, Dallas, Texas.
19. *Falls City Journal*, "Photos Tell Tragic Story of Airliner Crash," August 9, 1966, p. 6.
20. *Falls City Journal*, "Identification of Victims a Gruesome Job; Many See Crash," August 9, 1966, p. 2.
21. Ibid.
22. *Falls City Journal*, "Back to Normal Here After 12 Hectic Days," August 19, 1966, p. 1.
23. Braniff International Airways, Internal Document: Funeral Status Report 6a, 10:00 a.m., August 10, 1966, Braniff Airways Archives, Richardson, Texas, University of Texas-Dallas.
24. *Falls City Journal*, "Back to Normal Here," August 19, 1966, p. 1.
25. Steve Pollock interview with Kenny and Elaine Schawang, Falls City, Nebraska, October 26, 2012.
26. *Falls City Journal*, "42 Dead," p. 1.

Chapter 17

1. *Falls City Journal*, "Investigation Continues at Scene at Air Tragedy," August 9, 1966, p. 1.
2. Ibid.
3. Ibid., 2.
4. Ibid.
5. *Falls City Journal*, "Identification of Victims a Gruesome Job; Many See Crash," August 9, 1966, p. 2.
6. *Chicago Tribune*, "Hope 2d Tape Holds Clew in Plane Crash," August 9, 1966, p. B7.
7. United States Department of Transportation, Federal Aviation Administration, Regulation 14 CFR 125.227: Cockpit Voice Recorders.
8. *Chicago Tribune*, "Hope," August 9, 1966, p. B7.
9. *Syracuse Post-Standard*, "Plane Recorders to Be Made Mandatory," December 12, 1963, p. 4.
10. State of Illinois, 92nd General Assembly, Legislation, House Joint Resolution 87 (2002).
11. Ibid.
12. Department of Transportation, *The Federal Aviation Administration: A Historical Perspective, 1903–2008* (Washington: Department of Transportation, 2008), chapter 4.
13. *Lubbock Avalanche-Journal*, "Death Throes of Plane Included: Recording of Air Crash Revealed," December 8, 1966, p. B1.
14. Civil Aeronautics Board, Report on Braniff International Airways Flight 250, Report No. SA393-1-0008 (hereafter CAB BI250), 29, April 18, 1968.
15. *Lubbock Avalanche-Journal*, "Death Throes of Plane Included," December 8, 1966, p. B1.
16. CAB BI250, pp. 29–30.
17. See chapter 13.
18. Warren R. Young, *Life*, December 18, 1964, 86; see chapter 4.

Chapter 18

1. *Falls City Journal*, "Congressman Callan Has Been Riding the Ill-Fated Airliner," August 8, 1966, p. 1.
2. Tetsuya Theodore Fujita, "Memoirs of Effort to Unlock the Mystery of Severe Storms During the 50 Years 1942–1992," Texas Tech

University, Wind Research Laboratory Paper Number 239 (1994), chapter 1.
 3. Ibid.
 4. Ibid.
 5. Ibid.
 6. Ibid., chapter 5.
 7. Ibid., chapter 6.
 8. Ibid., chapter 2.
 9. *Lawrence* (KS) *Journal-World*, "Weather Blamed During Hearing on Jet Crackup," December 8, 1966, p. 2.
 10. Tetsuya Fujita, "Detailed Investigation of Mesometeorological Conditions of the Squall Line of August 6–7, 1966, which Crossed the Air Route between Kansas City, Missouri and Omaha, Nebraska, Part I, Text and Figures," for British Aircraft Corporation (U.S.A.) Inc., November 1966.
 11. Ibid., 27.
 12. Ibid., 28.
 13. Ibid., 29.
 14. Ibid.
 15. Ibid., 2.
 16. Ibid., 3.
 17. Fujita, "Detailed Investigation of Mesometeorological Conditions of the Squall Line of August 6–7, 1966."
 18. Ibid., 2.
 19. Ibid., 5.
 20. Ibid.
 21. Ibid., 6.
 22. Ibid., 7.
 23. Ibid., 11.
 24. Ibid., 27.
 25. Ibid., 29.
 26. James R. McDonald, "T. Theodore Fujita: His Contribution to Tornado Knowledge through Damage Documentation and the Fujita Scale," *Bulletin of the American Meteorological Society* 82, no. 1 (January 2001).

Chapter 19

 1. Civil Aeronautics Board, April 18, 1968, Report on Braniff International Airways Flight 250, Report No. SA393-1-0008 (hereafter CAB BI250), 18.
 2. Ibid.
 3. Ibid., 20.
 4. Ibid., 13.
 5. Ibid., 22–23.
 6. Ibid., 23.
 7. Ibid., 24.
 8. Ibid.
 9. Ibid., 24–25.

 10. Ibid., 25.
 11. Ibid.
 12. Ibid., 28–29.
 13. See chapter 19.
 14. CAB BI250, 31.
 15. Ibid., 34.
 16. Ibid., 5.
 17. Ibid., 43–45.
 18. Ibid., 53.
 19. Ibid., 54.
 20. Ibid., 59.
 21. Department of Transportation, The Federal Aviation Administration: A Historical Perspective, 1903–2008 (Washington, D.C.: Department of Transportation, 2008), chapter 8.
 22. *Columbus* (OH) *Daily Telegraph*, "84 Killed in Texas Plane Crash," May 4, 1968, p. 1.
 23. See chapter 11.

Chapter 20

 1. *Omaha World Herald*, "Probers Still Search for Clues to Air Crash," August 9, 1966, p. 1A.
 2. *Huron Daily Plainsman*, "Many of 42 Victims Were on Visits," August 8, 1966, p. 1.
 3. Ibid.
 4. *Falls City Journal*, "1966 Braniff Air Crash Never Out of Mind," April 28, 2006, p. 1.
 5. Ibid.
 6. Steve Pollock interview with Tim Hilliker, Atlanta, Georgia, March 30, 2013.
 7. Steve Pollock interview with Elaine and Kenny Schawang, Falls City, Nebraska, October 26, 2012.
 8. Memorial Marker, Braniff 250 Accident Site, 7.6 Miles NNE of Falls City, Nebraska, author visit, October 26, 2013.
 9. John 14:1–3, New Testament, New International Version.
 10. *Falls City Journal*, "1966 Braniff Air Crash Never Out of Mind," p. 2.
 11. Steve Pollock interview with Bill Schock, Falls City, Nebraska, October 26, 2013.
 12. Steve Pollock interview with Dan Hilliker, Eden Prairie, Minnesota, October 27, 2013.
 13. *Falls City Journal*, "1966 Braniff Air Crash Never Out of Mind," p. 2.
 14. Steve Pollock interviews with Dan Hilliker, Bob Lindseth, and Kathy Prestegard, Eden Prairie, Minnesota, October 27, 2013.
 15. Ibid.
 16. Ibid.

Bibliography

Books

Bechtel, Stefan, and Tim Samaras. *Tornado Hunter: Getting Inside the Most Violent Storms on Earth*. Des Moines: National Geographic Books, 2009.

Captain x and Reynolds Dodson. *Unfriendly Skies: Revelations of a Deregulated Airline Pilot*. New York: Doubleday, 1989.

Denham, Terry. *World Directory of Airliner Crashes*. Sparkford Near Yeovil, Somerset, England: Patrick Stephens, 1996.

Doka, Kenneth J., and Marcia E. Lattanzi-Licht, ed. *Coping with Public Tragedy (Living with Grief.)* Washington, D.C.: Hospice Foundation of America, 2003.

Fink, George, ed. *Stress of War, Conflict and Disaster*. Amsterdam: Academic, 2010.

Gandt, Robert. *Skygods: The Fall of Pan Am*. New York: William Morrow, 1995.

Gero, David. *Aviation Disasters: The World's Major Civil Airliner Crashes Since 1950*. 3rd ed. Sparkford Near Yeovil, Somerset, England: Patrick Stephens, 2000.

Gist, Richard, and Bernard Lubin. *Response to Disaster: Psychosocial, Community, and Ecological Approaches*. Series in Clinical and Community Psychology. New York: Routledge, 1999.

Hailey, Arthur. *Airport*. New York: Doubleday, 1968.

Hirschhorn, Clive. *Gene Kelly: A Biography*. New York: St. Martin's, 1985.

Hurt, H.H., Jr. *Aerodynamics for Naval Aviators*. Los Angeles: University of Southern California Press, Revised January 2, 1965.

Job, Macarthur. *Air Disaster*. Fyshwick, Australia: Aerospace, vol. 1, 1994; vol. 2, 1996; vol. 3, 1998.

_____. *Air Disaster*. Vol. 4: *The Propeller Era*. Fyshwick, Australia: Aerospace, 2001.

King, Adele. *Rereading Camara Laye*. Lincoln: University of Nebraska Press, 2002.

Koren, Jay. *The Company We Kept*. McLean, VA: Paladwr, 2000.

Kurnitz, Harry. *Reclining Figure*. New York: Dramatist's Play Service, 1998.

Lawrence, Mary Wells. *A Big Life in Advertising*. New York: Simon & Schuster, 2003.

MacPherson, Malcolm, ed. *The Black Box: All-New Cockpit Voice Recorder Accounts of In-Flight Accidents*. New York: Quill–William Morrow, 1998.

_____. *On a Wing and a Prayer: Interviews with Airline Disaster Survivors*. New York: Perennial, 2002.

Marriott, Leo, Stanley Stewart and Michael Sharpe. *Air Disasters: Including Dialogue from the Black Box*. London: PRC, 1999.

McClement, Fred. *Anvil of the Gods: Modern Airplanes Versus Violent Storms*. New York: Lippincott, 1964.

_____. *It Doesn't Matter Where You Sit; Air Disasters: Why and How They Happen, and the Menacing Future of Jet Travel*. New York: Holt, Rinehart and Winston, 1969.

Megson, T.H.G. *Introduction to Aircraft Structural Analysis*. Amsterdam: Elsevier, 2010.

Nance, John J. *Splash of Colors: The Self-Destruction of Braniff International*. New York: William Morrow, 1984.
Oliver, Susan. *Odyssey: A Daring Transatlantic Journey*. New York: Macmillan, 1983.
Ray, Sally J. *Strategic Communication in Crisis Management: Lessons from the Airline Industry*. Westport, CT: Quorum, 1999.
Robinson, Jack E. *Free Fall: The Needless Destruction of Eastern Air Lines and the Valiant Struggle to Save It*. New York: HarperBusiness, 1992.
Sampson, Anthony. *Empires of the Sky: The Politics, Contests, and Cartels of World Airlines*. New York: Random House, 1985.
Scaglia, Beatriz. *Deadliest Air: The Fifteen Worst Airline Accidents, Collisions or Acts of Terrorism (Pan Am Flight 103, Tenerife Airport Disaster, American Airlines Flight 587, and More)*. Baldwin City, KS: Webster's Digital Services, 2011.
_____. *Lost Athletic Souls: Notable Sporting Airline Disasters*. Baldwin City, KS: Webster's Digital Services, 2011.
Schock, Bill. *Thrills, Chills and a Spill*. Falls City, NE: Self-Published, 2010.
Serling, Robert J. *The Electra Story: Aviation's Greatest Mystery*. New York: Bantam, 1963.
Smith, Michael. *Warnings: The True Story of How Science Tamed the Weather*. Austin, TX: Greenleaf, 2010.
Szurovy, Geza. *Classic American Airlines*. Minneapolis: MBI, 2003.
Williams, Charles M. *Crash of TWA Flight 260*. Albuquerque: University of New Mexico Press, 2010.
Yenne, Bill. *Northwest Orient*. London: Bison, 1996.

Documents, Reports

Braniff International Airways Internal Document: Funeral Status Report 6a, 10:00 a.m., August 10, 1966. Richardson, TX: University of Texas–Dallas Braniff Airways Archives.
British Aircraft Corporation. BAC One-Eleven Sales Brochure, Richardson: University of Texas–Dallas, Braniff Airways Archives, 1964.
Bureau d'Enquêtes et d'Analyses pour la sécurité de l'aviation civilé Ministère de l'Écologie, du Développement durable, des Transports et du Logement (BEA). "On the Accident on 1st June 2009 to the Airbus A330–203 Registered F-GZCP Operated by Air France Flight AF447 Rio de Janeiro-Paris," July 28, 2012.
Civil Aeronautics Board. "Eastern Air Lines, Inc. Douglas DC-8, N8607, New Orleans, Louisiana, February 25, 1964." Report No. SA379/1–0006. Washington, D.C.: Civil Aeronautics Board, July 1, 1966.
_____. "The Flying Tiger Line Inc., L-1049H, N6915C, San Francisco International Airport, San Francisco, California, December 24, 1964." Report No. SA382/1–0064. Washington, D.C.: Civil Aeronautics Board, June 8, 1966.
_____. Northwest Airlines, Inc. "Boeing 720B, N724US, near Miami, Florida, February 12, 1963." Report No. SA372/1–0006. Washington, D.C.: Civil Aeronautics Board, June 4, 1965.
_____. "Pan American World Airways Boeing 707, N712PA, over the Atlantic between London, England, and Gander, Newfoundland, February 3, 1959." Report No. US COMM-DC-25158/1–0006. Washington, D.C.: Civil Aeronautics Board, November 3, 1959.
_____. "Pan American World Airways, Inc., Boeing 707–121, N709PA, Near Elkton, Maryland, December 8, 1963." Report No. SA-376/1–0015. Washington, D.C.: Civil Aeronautics Board, March 3, 1965.
Conseil National ces Femmes Françaises. "Marie-Hélène Lefaucheux, 26 Février 1904–25 Février 1964." Obituary 1964.
Hoover, J. Edgar. Director, Federal Bureau of Investigation, United States Department of Justice. Letter of Sept. 26, 1966, to Walter Henshel, Vice President of Public Relations, Braniff International Airways, Dallas, Texas.
International Civil Aviation Organization. "Reports on Air Transportation in 1963." Press Release. Montreal, Canada, May 1, 1964.
National Air Transportation Association. "Aircraft Maintenance and System Tech-

nology Committee Best Practices: Minimum Equipment List (MEL)." Alexandria, VA: National Air Transportation Association, Fall 2012.

National Transportation Safety Board. Aircraft Accident Report, Alaska Airlines Flight 261, McDonnell Douglas MD-83, N963AS, About 2.7 Miles North of Anacapa Island, California, January 31, 2000. Report No. NTSB AAR-02/01/PB2002-910402, "Loss of Control and Impact with Pacific Ocean." Washington, D.C.: National Transportation Safety Board, December 30, 2002.

———. Aircraft Accident Report, Allegheny Airlines, Inc., Douglas DC-9, N994VJ, at Philadelphia, Pennsylvania, June 23, 1976. Report No. NTSB-AAR-78-2. Washington, D.C.: National Transportation Safety Board, January 18, 1978.

———. Aircraft Accident Report, American Airlines Flight 587, Airbus Industrie A300-605R, N14053, Belle Harbor, New York, November 12, 2001. Report No. NTSB/AAR-04/04/PB2004-910404/Notation 7439B, "In-Flight Separation of Vertical Stabilizer." Washington, D.C.: National Transportation Safety Board, 2004.

———. Aircraft Accident Report, Braniff Airways, Inc., BAC 1-11, N1553, near Falls City, Nebraska, August 6, 1966. Report No. SA393/1-0008. Washington, D.C.: National Transportation Safety Board, April 18, 1968.

———. Aircraft Accident Report, Braniff Airways, Inc. Lockheed L-188 N9707C, near Dawson, TX, May 3, 1968. Report No. SA403/1-0003. Washington, D.C.: National Transportation Safety Board, June 19, 1969.

———. Aircraft Accident Report, Continental Air Lines, Inc., Boeing 727-224, N88777, Stapleton International Airport, Denver, Colorado, August 7, 1975. Report No. NTSB-AAR-76-14. Washington, D.C.: National Transportation Safety Board, May 5, 1976.

———. Aircraft Accident Report, Continental Air Lines, Inc., Boeing 727-224, N32725, Tucson, Arizona, June 3, 1977. Report No. NTSB-AAR-78-9. Washington, D.C.: National Transportation Safety Board, August 1, 1978.

———. Aircraft Accident Report, Delta Air Lines, Inc., Douglas DC-8, N802E, Kenner, Louisiana, March 30, 1967. Report No. SA397/1-0003. Washington, D.C.: National Transportation Safety Board, December 20, 1967.

———. Aircraft Accident Report, Delta Air Lines, Inc., Lockheed L-1011-385-1, N726DA, Dallas/Fort Worth International Airport, Texas, August 2, 1985. Report No. NTSB-AAR-86-05. Washington, D.C.: National Transportation Safety Board, August 15, 1986.

———. Aircraft Accident Report, Eastern Air Lines, Inc., Boeing 727-25 N8139, Atlanta Hartsfield International Airport, Atlanta, Georgia, August 22, 1979. Report No. NTSB-AAR-80-6. Washington, D.C.: National Transportation Safety Board, May 28, 1980.

———. Aircraft Accident Report, Eastern Air Lines, Inc., Boeing 727-225, John F. Kennedy International Airport, Jamaica, New York, June 24, 1975. Report No. NTSB-AAR-76-8. Washington, D.C.: National Transportation Safety Board, March 12, 1976.

———. Aircraft Accident Report, Eastern Air Lines, Inc., Douglas DC-9-31, N8984E, Charlotte, North Carolina, September 11, 1974. Report No. NTSB-AAR-75-9/1-0020. Washington, D.C.: National Transportation Safety Board, May 23, 1975.

———. Aircraft Accident Report, Loftleidir Icelandic Airlines, Inc., Douglas DC-8-61, John F. Kennedy International Airport, Jamaica, New York, June 23, 1973. Report No. NTSB-AAR-73-20/A-0003. Washington, D.C.: National Transportation Safety Board, December 5, 1973.

———. "Safety Recommendation Regarding American Airlines Flight 587." Report A-10-119 and -120, A-04-63. Washington, D.C.: National Transportation Safety Board, August 4, 2010.

———. Aircraft Accident Report, "Uncontrolled Collision with Terrain Flagship Airlines, Inc., dba American Eagle, Flight

5379, BAe Jetstream 2201, N918AE, Morrisville, North Carolina December 13, 1994." Report No. NTSB-AAR-95/07/DCA95MA006. Washington, D.C.: National Transportation Safety Board, October 24, 1995.

Hoyert, Donna L., and Jiaquan Xu. "Deaths: Preliminary Data for 2011." *National Vital Statistics Reports* 61, no. 6 (October 10, 2012). Hyattsville, MD: U.S. Dept. of Health and Human Services.

PPG. "Facts About PPG." Company brochure, 2012.

United Nations. "Short History of the Commission on the Status of Women." 2000.

U.S. Army Air Force. "Initial Selection of Candidates for Pilot, Bombardier and Navigator Training." "Prepared by Assistant Chief of Air Staff Intelligence, Historical Division." *Army Air Forces Historical Studies*, no. 2 (November 1943).

U.S. Department of Transportation, Federal Aviation Administration. Advisory Circular: Atmospheric Turbulence Avoidance. AC No. 00-30B. Washington, D.C., September 9, 1997.

_____. Advisory Circular: Pilot Windshear Guide. AC No. 00-54. Washington, D.C., November 25, 1988.

_____. Advisory Circular: Preventing Injuries Caused by Turbulence. AC No. 120-88A. Washington, D.C., January 19, 2006.

_____. Advisory Circular: Thunderstorms. AC No. 00-24B. Washington, D.C., January 20, 1983.

_____. Advisory Circular: Use of Portable Electronic Devices Aboard Aircraft. AC No. 91-21.1B. Washington, D.C., August 25, 2006.

_____. Aeronautical Information Manual. *Official Guide to Basic Flight Information and ATC Procedures*. Washington, D.C.: February 9, 2012.

_____. Flight Standards Service. *Pilot's Handbook of Aeronautical Knowledge*. FAA H-8083-25A. Washington, D.C., 2008.

United States Statutes at Large. Civil Aeronautics Act of 1938, p. 973. Washington D.C.: United States Government Printing Office.

Magazines

Airways Classics. "Braniff." Special Edition, 2010.

Angell, Roger. "Life During Wartime: New York, 1967." *New Yorker*, June 12, 2006, p. 22.

Business Week. "Autopsy of an Airliner: How CAB Operates; Last Week's Crash of a Braniff Jet Brought Government and Industry Investigators to Nebraska to Search for 'Probable Cause.'" August 13, 1966, Transportation Section, pp. 119–123.

Flight. "DC-8." November 18, 1960, p. 803.

"Flight Crew of DC-10 Encounters Microburst During Unstabilized Approach, Ending in Runway Accident." *Flight Safety Foundation Accident Prevention* 53, no. 8 (August 1996), 1.

Miller, Carol. "While a Jet Fell 2½ miles 'I was on the ceiling.'" *Life*, December 6, 1963, p. 23.

Rosenkrans, Wayne. "Not Worth Being Upset: Recent U.S. Law Influences Specialists' Proposals for Simulator Upgrades and Limited Use of All-Attitude, All-Envelope Training Airplanes." Flight Safety Foundation. *AeroSafetyWorld*, June 2011, p. 24.

Tinker, Frank A. "Diaries of Destruction: Will Their Secrets Make Flying Safer?" *Popular Mechanics*, June 1963, p. 103.

Young, Warren R. "Turbulence: Hidden Giant in the Sky." *Life*, December 18, 1964, p. 86.

Scientific Papers

Civil Aviation Authority Safety Regulation Group. "A Benefit Analysis for Cabin Water Spray Systems and Enhanced Fuselage Burnthrough Protection," CAA Paper 2002/04.

FAA reference: DOT/FFA/AR-02/49. RGW Cherry and Associates, April 7, 2003.

Fielding, E., W. Lo and J.H. Yang. "The National Transportation Safety Board: A Model for Systemic Risk Management." *Journal of Investment Management* 9 (2011): 18–50.

Fujita, Tetsuya. "Detailed investigation of mesometeorological conditions of the squall line of August 6–7, 1966 which crossed the air route between Kansas City, Missouri and Omaha, Nebraska: Part I, Text and Figures; Part II, References—Reproductions; Part III, Turbulence in relation to the squall line." Prepared for British Aircraft Corporation (U.S.A.), Inc. Urbana: University of Chicago. November, 1966.

Fujita, Tetsuya, Dorothy L. Bradbury and C.F. Van Thullenar. "Palm Sunday Tornadoes of April 11, 1965." *Monthly Weather Review* 98, no. 1 (January 1970): 29–69.

Fujita, T. Theodore, and Horace R. Byers. "Spearhead Echo and Downburst in the Crash of an Airliner." *Monthly Weather Review* 105, no. 2 (February 1977).

Fujita, T. Theodore. "Proposed Characterization of Tornadoes and Hurricanes by Area and Intensity." University of Chicago, Satellite and Mesometeorology Research Project, Department of the Geophysical Sciences. Unclassified SMRP Research Paper Number 91, February 1971.

Hess, R.A. "Unified Theory for Aircraft Handling Qualities and Adverse Aircraft-pilot Coupling." *Journal of Guidance, Control, and Dynamics* 20, no. 6 (1997): 1141–1148.

Klemp, Joseph. "Dynamics of Tornadic Thunderstorms." National Center for Atmospheric Research, Boulder, Colorado. Annual Review of Fluid Mechanics, 1987.

Loeffler, Andrew R. "'It Is Now Safe to Move About the Cabin': Revisiting Air Carrier Liability for Passenger Injuries Due to Turbulence Using a Res Ipsa Loquitur Theory of Negligence." *Northern Illinois University Law Review* 30 (May 17, 2010): 445–480.

McDonald, James R. "T. Theodore Fujita: His Contribution to Tornado Knowledge through Damage Documentation and the Fujita Scale." *Bulletin of the American Meteorological Society* 82, no. 1 (January 2001).

McDonough, J.M. "Introductory Lectures on Turbulence: Physics, Mathematics and Modeling." Departments of Mechanical Engineering and Mathematics, University of Kentucky, 2004, 2007.

National Aeronautics and Space Administration. "NASA Conference on 'Some Problems Related to Aircraft Operation': Compilation of the Papers Presented." Langley Research Center, Langley Field, VA, November 5–6, 1958. N65–88161.

Proctor, Fred H., David W. Hamilton, David K. Rutishauser and George F. Switzer. "Meteorology and Wake Vortex Influence on the American Airlines FL-587 Accident." Langley Research Center and Research Triangle Institute, Hampton, VA, 2004. NASA/TM-2004–213018.

Ragland, S., Jr., R.M. Chambers, R.J. Crosbie and L. Hitchcock, Jr. "Simulation and Effects of Severe Turbulence on Jet Airline Pilots." NADC-ML-64 11. Johnsville, PA, August 13, 1964.

Steiner, Roy. "An Analysis of Normal Accelerations and Airspeeds of One Type of Twin-Engine Transport Airplane in Commercial Operations Over a Northern Transcontinental Route." Langley Aeronautical Laboratory, Langley Field, VA. Technical Note 2833. Washington, D.C.: National Advisory Committee for Aeronautics, November 1952.

Steiner, Roy, and Martin R. Copp. "A Review of Atmospheric Turbulence and Its Significance to Jet-Transport Operations." Langley Research Center (later a part of NASA). Presented at NASA's Conference "Some Problems Related to Aircraft Operation," 1958.

Tiemeyer, Philip James. "Manhood Up in the Air: Gender, Sexuality, Corporate Culture, and the Law in Twentieth Century America." Ph.D. diss., University of Texas at Austin, May 2007.

Journals

"British Aircraft Corporation One-Eleven." *Aircraft Engineering Journal* 35, no. 5 (September 1964), 1–32.

Locke, Brian. "Strange Fruit: White, Black, and Asian in the WWII Combat Film Bataan." *Journal of Popular Film and Television* 36, no. 1 (Spring 2008), 9–20.

Newspaper Articles, by subject

1966

Cochran, Mike (AP). "Convicted Killer McDuff: Portrait of Brutal Rage: 26 Years After 3 Teens Lost Their Lives to this Misfit, Police Say He Has Struck Again and Again." *Kerrville (TX) Daily Times*, May 18, 1992, P. 6A.

Haines, Max. "Born to Raise Hell." Crime Flashback, *Lethbridge (AB) Herald*, June 5, 2005, P. A8.

"Mao: China's Great Leader Put His Faith in Peasants." *Brandon (AB) Sun*, September 11, 1976P. 5.

Pett, Saul, and Jules Loh. "Trail Through the Sniper's Mind: What Made Charley Whitman Tick?" *San Antonio Sunday Express and News*, August 7, 1966, Section D, P. 1.

Braniff Flight 250

"Airline Is One Man's Story; Braniff is First Airline in Laredo in 15 Years; First Flight to Arrive Sunday." *Laredo Times* (TX), Aug. 29, 1943, P. 1.

"Airliner Crash Victims Listed." *Albuquerque Journal*, Aug. 8, 1966, P. 2.

"All 42 Die as Jet Falls in Flames in Nebraska Field; Witnesses Tell of Flash and Explosion Before Braniff Airliner Hit the Ground; Plane Built in Britain; U.S. Investigators to Study Wreck of First BAC 111 to Have Accident in Nation." *New York Times*, Aug. 8, 1966, P. 1.

"All Perish: Jetliner Falls With 42." *San Antonio Light* (TX), Aug. 8, 1966, P. 2.

"Back to Normal Here After 12 Hectic Days." *Falls City Journal* (NE), Aug. 19, 1966, P. 1.

Beckler, John, "Negotiators Resume Air Strike Talks." *Huron Daily Plainsman* (SD), Aug. 8, 1966, P. 1.

"Being Removed Crash Wreckage; Memorial Services." *Falls City Journal* (NE), Aug. 15, 1966, P. 1.

"Big Roles for Civil Defense, Guards." *Falls City Journal* (NE), Aug. 8, 1966, P. 1.

"Braniff Applies for Transpacific Route." *Oakland Tribune*, Dec. 10, 1966, P. 28.

"Braniff Asks for Round the World Route." *Colorado Springs Gazette-Telegraph*, Feb. 27, 1966, P. 5E.

"Braniff in the Air Over New Global Fashion; Leotards, Shifts, Fur Coats Become Uniform." *Grand Prairie Daily News* (TX), Aug. 8, 1966, P. 7.

"Braniff Jetliner Falls Into Field on Schawang Farm; Many Persons See Tragedy Develop Aboard Airliner Headed for Omaha; Officials Here to Investigate the Crash." *Falls City Journal* (NE), Aug. 8, 1966, P. 1.

"CAB Probes Air Crash Fatal to 42." *Salt Lake Tribune*, Aug. 8, 1966, P. 1.

"CAB Seeks Cause of Airline Crash; Recording May Hold Vital Clue." *Jacksonville Courier* (IL), Aug. 8, 1966, P. 1.

Carmichael, Dan. "Is American Airlines Bankrupting Braniff International 'Dirty Tricks?'" *Elyria Chronicle-Telegram* (OH), Mar. 11, 1982, P. A-4.

Clark, Evert. "Turbulence Role in Crash Studied; Nebraska Disaster Reopens Question of Violent Air." *New York Times*, Aug. 21, 1966, P. 30.

"Chic Airline Hostesses Take Off with New Look." *Long Beach Press-Telegram*, Jul. 26, 1964, P. B-6.

"Clues Sought to Jet Crash in Nebraska." *Huron Daily Plainsman* (SD), Aug. 8, 1966, P. 1.

"Colorful Braniff Jet Lands Here." *Oakland Tribune*, Jun. 5, 1966, P. 34.

"Congressman Callan Has Been Riding the Ill-Fated Airliner." *Falls City Journal* (NE), Aug. 8, 1966, P. 1.

"Death Throes of Plane Included: Recording of Air Crash Revealed." *Lubbock Avalanche-Journal* (TX), Dec. 8, 1966, P. B1.

"Experience a Terrible Nightmare." *Falls City Journal* (NE), Aug. 8, 1966, P. 1.

"FAA Cites Ex-Lawmaker's Push for Black Boxes." *Chicago Tribune*, Dec. 19, 1998, P. 1.

"Fiery Crash of Twin-Jet Craft Kills 42; D.C. Woman Among Victims; Nebraska Wreckage Studied." *Washington Post*, Aug. 8, 1966, P. A1.

"50 Troops Brought in to Assist in Search; Area Along Flight Path of Jetliner Will Be Combed for Pieces; Ask Farmer's

Help." *Falls City Journal* (NE), Aug. 11, 1966, P. 1.
"Flight Recorder Ruined by Heat in Airline Crash." *Stevens Point Daily Journal* (WI), Aug. 8, 1966, P. 9.
"Flight Recorder Ruined by Heat in Airline Crash." *Wisconsin Daily Journal* (Madison), Aug. 8, 1966, P. 9.
"42 Dead in Airliner Crash; Crash Scene Ghastly Sight." *Falls City Journal* (NE), Aug. 8, 1966, P. 1.
"Fuel Link Hinted in Crash of Jet; Rough Weather Also Cited—Accident Fatal to 42." *New York Times*, Aug. 9, 1966, P. 16.
"Hearing on Braniff Crash May Be Held in Falls City: Formal Hearings Must be Conducted in State Plane Tragedies Occur; Plane Crash Probe Next Month." *Falls City Journal* (NE), Sept. 22, 1966, P. 1.
"Hope 2d Tape Holds Clue in Plane Crash." *Chicago Tribune*, Aug. 9, 1966, P. B7.
"Identification of 42 Plane Crash Victims is Not As Yet Completed; Efficient Handling of Tragedy Praised by Gov. Morrison." *Falls City Journal* (NE), Aug. 9, 1966, P. 1.
"Identification of Victims a Gruesome Job; Many See Crash." *Falls City Journal* (NE), Aug. 9, 1966, P. 2.
"Investigation Continues at Scene at Air Tragedy." *Falls City Journal* (NE), Aug. 9, 1966, P. 1.
"Last Load of Air Wreckage on Way." *Falls City Journal* (NE), Aug. 18, 1966, P. 1.
"List of 42 Crash Victims." *New York Times*, Aug. 8, 1966, P. 24.
"List of Those on Ill-Fated Jet; 38 Passengers and Crew of 4 Killed." *Chicago Tribune*, Aug. 8, 1966, P. 10.
"Many of 42 Victims Were on Visits." *Huron Daily Plainsman* (SD), Aug. 8, 1966, P. 1.
"Many Persons Help in County Tragedy." *Falls City Journal* (NE), Aug. 8, 1966, P. 1.
Merzer, Martin. Knight-Ridder Newspapers. "Braniff is First Airline to File Reorganization Petition." *Orange County Register*, May 14, 1982, Business Section, P. D13.
"More Than Usual: Braniff Admits Some Flights Are Late." *Corpus Christi Times*, Dec. 23, 1966, P. 10.
"Mother, Son Among Dead." *Burlington Hawk-Eye* (IA), Aug. 8, 1966, P. 2.

"Moving Ahead with Wreckage Removal." *Falls City Journal* (NE), Aug. 17, 1966, P. 1.
"1966 Braniff Air Crash Never Out of Mind." *Falls City Journal* (NE), Apr. 28, 2006, P. 1.
"Photos Tell Tragic Story of Airliner Crash." *Falls City Journal* (NE), Aug. 9, 1966, P. 6.
"Plane Fuel Leak Hinted Crash Cause." *Washington Post*, Aug. 8, 1966, P. A3.
"Prince Rates 'Excellent' For His Flying Ability." *High Point Enterprise*, Mar. 13, 1966, P. 5A.
"Probers Still Search for Clues to Air Crash." *Omaha World Herald*, Aug. 9, 1966, P. 1A.
"Ready at the Crash Site." *Falls City Journal* (NE), Aug. 12, 1966, P. 1.
"Recorder Burned On Crashed Plane; 42 Killed In Crash In Nebraska." *Waterloo Daily Courier* (IA), Aug. 8, 1966, P. 8.
"Recording Device May Give Answer." *Chicago Tribune*, Aug. 8, 1966, P. 2.
"Removing Engines from the Jetliner Wreckage; Probes Pay Off." *Falls City Journal* (NE), Aug. 12, 1966, P. 2.
Satterwhite, Geraldine. "Mother Set Flight Pattern for Sherry's Braniff Wings." *Abilene Reporter-News* (TX), Oct. 29, 1966, P. 7-B.
"Senate Support for 2,000 MPH Airliner." *Falls City Journal* (NE), Aug. 11, 1966, P. 2.
Singer, Phyllis. "Shari Meets a Real Prince." *Waterloo Daily Courier* (IA), Mar. 15, 1966, P. 8.
"Some Reluctance: House Eyes Air Strike Action." *San Antonio Light* (TX), Aug. 8, 1966, P. 2.
"'Swirl of Wind' Heard on Plane's Recorder." *Falls City Journal* (NE), Sept. 12, 1966, P. 1.
Thomas, Wayne. "Speed Surge of Doomed Jet Revealed; Velocity Linked to Crash Fatal to 42 Aboard." *Chicago Tribune*, Aug. 11, 1966, P. A3.
"2 Recorders Hold Clues to Air Crash; Probe Debris and 42 Deaths." *Chicago Tribune*, Aug. 8, 1966, P. 1.
"$251,000 Suit Filed in Braniff Crash." *Falls City Journal* (NE), Aug. 24, 1966, P. 1.
"Weather Blamed During Hearing on Jet

Crackup. Dec. 8, 1966." *Lawrence Journal-World* (KS), P. 2.

"Weather May Have Figured in Jetliner Crash; Crash Like Daylight." *Falls City Journal* (NE), Aug. 10, 1966, P. 1.

"White House Okays Purchase of Panagra by Braniff Airways." *Bridgeport Post* (IA), Oct. 19, 1966, P. 66.

"Winding-Up Process at Braniff Crash Site." *Falls City Journal* (NE), Aug. 15, 1966, P. 1.

"Witness to Nebraska Crash: Airliner Dropped as 'Ball of Flames.'" *Salt Lake Tribune*, Aug. 8, 1966, P. 5.

Braniff Flight 352

"84 Killed in Texas Plane Crash." *Columbus Daily Telegraph* (OH), May 4, 1968, P. 1.

Capital Airlines Flight 75

"Capital Airlines Crashes Take Lives of 33 Persons; Turboprop Explodes, Killing 31." *Jefferson City Daily Capital News*, May 13, 1959, P. 1, Morning Edition.

Capital Airlines Flight 983

"Capital Airlines Crashes Take Lives of 33 Persons; Crash After Landing; Two Killed as Plane Plunges Over Bank." *Jefferson City Daily Capital News*, May 13, 1959, P. 1, Morning Edition.

Eastern Air Lines

"Autogiro Lands on South Lawn of White House; Feat is Part of Ceremony with Trophy." *Daily Northwestern* (IL), Apr. 22, 1931, P. 1.

"Aviation Pioneer is Suicide." *San Antonio Express and News*, Apr. 24, 1960, P. 15D.

"From Slow Start to Supersonic." *San Antonio Light* (TX), May 15, 1977, P. 9.

"Hub Society Woman's Sea Death Probed: Mrs. John Pitcairn Fell Overboard from Schooner." *Lowell Sun* (MA), Apr. 22, 1949, P. 1.

Eastern Flights 301/304

"Air Forces Call Platteville Man." *Wisconsin State Journal* (Madison), Feb. 15, 1943, P. 14.

"Bass-Baritone Coming to Ogden Tuesday." *Ogden Standard-Examiner* (UT), Oct. 21, 1945, P. 10B.

"Bits of Nothing." *Fayette County Leader* (IA), Mar. 23, 1939, P. 4.

"Board Caps Crash Probe." *Laurel Leader-Call* (MS), Jul. 18, 1964, P. 2.

"Boy Scouts to Plant Trees as Memorials." *Wisconsin State Journal* (Madison), May 27, 1946, State Page, P. 12.

"15 Women Experts Represent America, Europe in Assembly." *Abilene Reporter-News* (TX), Jan. 12, 1953, P. 4A.

"58 Perish in Crash of Airliner." *Bakersfield Californian*, Feb. 25, 1964, P. 1.

"58 Perish in New Orleans Air Crash." *Biloxi Daily Herald*, Feb. 25, 1964, P. 1.

"French Head of Women's Group Accuses U.N. of Nationalism "; "Frenchwomen Preside Over Two Meetings." *Winnipeg Free Press*, Jun. 22, 1957, P. 12.

"Helpless Women Who Were Nazi Victims Now Need Assistance." *Greeley Daily Tribune* (CO), May 18, 1950, P. 16.

"Jet Co-Pilot State Native." *Manitowoc Herald-Times* (WI), Feb. 27, 1964, P. 14-M.

"Jet Engine Found Near Houston." *San Antonio Express* (TX), Nov. 12, 1963, P. 1.

"Jet Loses Engine But Lands Safely." *News and Courier* (Charleston, SC), Nov. 10, 1963, P. 10-A.

"Jetliner Crash Kills 58." *Pacific Stars and Stripes*, Feb. 27, 1964, P. 1.

"Ken Spencer to Sing in Recital at High School." *Mason City Globe-Gazette* (IA), Apr. 27, 1940, P. 5.

"Kenneth Spencer is Coming Back May 8." *Fayette County Leader* (IA), Apr. 11, 1940, P. 1.

"Kenneth Spencer, Negro Singer, Giving Fourth Recital Here." *Winnipeg Free Press*, Oct. 19, 1940, P. 28.

"Kenneth Spencer Sings at College Sunday Evening." *Dunkirk Evening Observer* (NY), May 4, 1946, P. 13.

"Kenneth Spencer Sings Here Tonight." *Fayette County Leader* (IA), Mar. 16, 1939, P. 1.

"Kenneth Spencer Thrills Fayette." *Fayette County Leader* (IA), Mar. 23, 1939, P. 8.

"Leaders in U.N.'s Fight for Women." *Sioux Center News* (IA), Jun. 15, 1950, P. 1.

"Lieut. Grant Newby, Platteville, Gets Oak

Leaf Cluster." *Wisconsin State Journal* (Madison), Jul. 6, 1944, State Page, P. 15.

"Lieut. Grant Newby Undergoes Surgery." *Wisconsin State Journal* (Madison), Dec. 8, 1943, State Page, P. 11.

"Mme. Marie-Helene Lefaucheax [sic] (photo)." *Winnipeg Free Press*, Feb. 26, 1964, P. 3.

"Missions Over, Newby Home." *Wisconsin State Journal* (Madison), Mar. 9, 1945, P. 7.

"More Bodies of Plane Crash Victims Found." *Altoona Mirror* (PA), Mar. 19, 1964, P. 22.

"Mother's Rights." *Lawton Constitution* (OK), Jul. 1, 1963, P. 15.

"Mystery Surrounds Jet Death Plunge." *Billings Gazette* (MT), Feb. 26, 1964, P. 1.

"The Nation ... A Wauwatosa couple ...." *Milwaukee Sentinel*, Feb. 26, 1964, P. 1.

"New Orleans Crash Leaves No Survivors." *Findlay Republican-Courier* (OH), Feb. 26, 1964, P. 1.

"Newby Promoted to Pilot." *Wisconsin State Journal* (Madison), Aug. 18, 1944, State Page, P. 7.

"Newby Wins Commission." *Wisconsin State Journal* (Madison), Nov. 3, 1943, State Page, P. 10.

"No Trace of 58 Lost in Lake Plane Crash." *Utica Daily Press*, Feb. 26, 1964, P. 1.

"Noted Soloist, Kenneth Spencer." *Berkeley Daily Gazette*, Mar. 4, 1939, P. 10.

"128 Escape Injury—Probers Seek Cause in Airliner Mishap." *Charleston Daily Mail* (SC), Nov. 11, 1963, P. 2.

"128 Safe as Crippled Airliner Lands After Downdraft Plunge." *Schenectady Gazette*, Nov. 11, 1963, P. 2.

"Pilot of Doomed Jetliner Radioed No Indication That Trouble Loomed." *Racine Journal-Times* (WI), Feb. 26, 1964, P. 16A.

"Platteville Men Train in Service." *Wisconsin State Journal* (Madison), Apr. 8, 1943, State Page, P. 11.

"Pontchartrain Lake Probed for Plane Wreckage." *Hagerstown Daily Mail* (MD), Mar. 14, 1964, P. 1.

"Remnants of Clothing and Big Silent Lake." *Laurel Leader-Call* (MS), Feb. 26, 1964, P. 3.

"Searchers Reach Wrecked Jetliner." *Oakland Tribune*, Mar. 15, 1964, P. 1.

"Seek to Aid Victims of Nazi Experiments." *Chicago Heights Star*, Dec. 11, 1951, P. 17.

"Serving Overseas, They Get Good News." *Wisconsin State Journal* (Madison), Jul. 16, 1944, State Page, P. 17.

"Six Passengers Injured; Jetliner Loses Engine But Lands Safely." *Lowell Sun* (MA), Nov. 10, 1963, P. 1.

"Speed Search for Jetliner Wreckage." *Lowell Sun* (MA), Feb. 26, 1964, P. 30.

"Steered into Thunderstorm, Jetliner Pilot is Helpless." Herald-Tribune News Service, *Hutchinson News* (KS), Nov. 21, 1963, P. 28.

"'Strange Sight, Weird Thing'—Stewardess Recalls Jetliner Accident." *Pampa Daily News* (TX), Nov. 11, 63, P. 1.

"Threat Halts Plane." *Pasadena Independent*, Nov. 13, 1963, P. 2.

"26 Platteville Students Get High Rankings." *Madison Capital Times* (WI), Oct. 22, 1941, P. 10.

"2 Tar Heels Die in Crash of Jetliner"; "La. Jetliner Crash Claims Lives of 58." *Wilmington Morning Star*, Feb. 26, 1964, P. 1.

"UN Women to Tackle One of Toughest Jobs." *Waterloo Daily Courier* (IA), Feb. 6, 1947, P. 11.

"Women Debate on Equal Footing with Men in United Nations." *Dothan Eagle* (AL), Dec. 4, 1947, P. 9.

"Women Experts Represent America and Europe in UN." *Harrisonburg Daily News-Record* (VA), Jan. 9, 1953, P. 13.

"Women's Groups Offer Contrasts in Gatherings." *Anderson Herald* (IN), Jun. 27, 1963, P. 19.

Northwest Flight 705

"Airliner Missing with 40: Ship Bound from Miami to Chicago." *Cedar Rapids Gazette*, Feb. 12, 1963, P. 1.

Bishop, Jim. "Reporter: True Crash Mystery." *San Antonio Light* (TX), May 17, 1963, P. 53.

"CAB Says Airliner's Design Contributed to Crash." *Corpus Christi Times*, Jun. 4, 1965, P. 1, Evening Edition.

"FBI, Aeronautics Board Probers Work in Secret Seeking Cause of Disasterous [sic] Crash in Everglades; 43 Persons Die as

Plane Plummets into Swamp After Takeoff from Miami Airport." *Ada Evening News* (OK), Feb. 13, 1963, P. 1.

"Flight 705's Last Minutes; And Long Hours of Waiting; 'Fireball Burst in Thunderhead.'" *San Mateo Times* (CA), Feb. 13, 1963, P. 4.

"Hint Mid-Air Breakup of Plane; 43 Die; Liner Wreckage Scattered Over Two-Mile Area; Daughter Weeps on Learning of Fatal Crash." *Long Beach Press-Telegram*, Feb. 13, 1963, P. 1.

"Jet in Mystery Crash; 43 Die: Down in Florida Swamp." *Valley Independent* (Monessen, Charleroi, Donora, PA), Feb. 13, 1963, P. 1.

"No Survivors: 43 Killed When Plane Crashes in Florida Swamp." *Connellsville Daily Courier* (PA), Feb. 13, 1963, P. 1.

"Seek Cause of Airliner Crash; 43 Die in Everglades Disaster; Chicago-Bound Plane Falls in Flame Minutes from Miami; Turn Seminole School into Morgue." *Mt. Vernon Register-News* (IL), Feb. 13, 1963, P. 1.

"Seek Cause of Florida Jet Crash: Continue Hunt for Bodies of All 43 Victims." *Racine Journal-Times* (WI), Feb. 13, 1963, P. 1A.

Pan American World Airways Flight 115

"Alert Crews Avert Two Air Mishaps." *Logansport Pharos-Tribune* (IN), Feb. 4, 1959, P. 10.

"B47 Crash Fatal to 3." *Elyria Chronicle-Telegram* (OH), Feb. 4, 1959, P. 1.

"Death Runs Amuck in Tuesday Skies," *Ada Evening News* (OK), Feb. 4, 1959, P. 2.

"Family Weekly Preview—Gene Kelly Tells of Near Jet Flight Crash." *Waterloo Daily Courier* (IA), Apr. 1, 1959, P.

"5 in Family on Plane, Boy is Only Survivor." *Elyria Chronicle-Telegram* (OH), Feb. 4, 1959, P. 1.

"How It Feels to Get 30,000-Foot Dive Aboard Jet." Cedar Rapids *Gazette* (IA), Feb. 4, 1959, P. 36.

Lyons, Leonard. "Robeson a Soviet Consultant"; "Gene Kelly and Susan Oliver were persuaded ...." *Middletown Daily Record* (NY), Feb. 13, 1959, P. 35.

"Show Goes On After 3 Stars Die in Crash." *Elyria Chronicle-Telegram* (OH), Feb. 4, 1959, P. 2.

"65 Believed Dead in Airliner Crash: Big Plane Plunges into River; Jet-Powered Craft Carrying 73 Misses New York Runway." *Elyria Chronicle-Telegram* (OH), Feb. 4, 1959, P. 1

"Tuesday was Tragic Day for Aviation." *Bennington Evening Banner* (VT), Feb. 5, 1959, P. 13.

"2 Jets Land Safely After Trouble in Air." *Elyria Chronicle-Telegram* (OH), Feb. 4, 1959, P. 2.

United Flight 746

"Duescher Awarded 10-Year Pin." *Roselle Register* (IL), Nov. 9, 1951, P1.

"Rain, Hail, Tornadoes Sweep U.S. Midlands." *Pasadena Independent Star-News* (UPI), Jul. 14, 1963, P. 3B.

"Turbulent Air Causes Airliner to Nosedive." *Bakersfield Californian*, Jul. 13, 1963, P. 26.

Index

Acker, C. Edward 102
Ackerman, Renate 96
Adams, Samuel Trustin 96
The Adventures of Ozzie and Harriet 18
AeroMexico 58
Aerospatiale Caravelle 180
Africa 77
Aiken, E.P. 48
Air Disaster (Volume I) 125
Air France 106
Air Medal with Oak Leaf cluster 74
Air Route Traffic Control Center 27; *see also* ARTCC
The Air Strip 111, 130
Airbus 85
Airbus A-380 86
Aircraft Accident Report 18, 42, 96; AAR 1-0006 18, 42; AAR SA 379 96
Aircraft Engineering Journal magazine 113, 120–122
aircraft nose down (AND) 88, 94–95
aircraft nose up (ANU) 88
Aircraft-Pilot Coupling (APC) 91
Airline Data Project 116
Airline Deregulation Act of 1978 100
Airline Pilots Association 161, 183
Airport 70
Akron, OH 29
Alaska Airlines 88; flight 261 88; MD-83 88
Alexandria, VA 97
All-School Parents' Day 71
Allen, Bobbie R. 44–45

Allen, Fred, Cpl. 146
Allentown, PA 97
Allies (World War II) 78–79
Allsup, Tommy 10
Almquist, Roy 23–25, 36, 39–42
AMAL *see* Aviation Medical Acceleration Laboratory
Amarillo, TX 101, 122
American Airlines 9–10, 63, 99, 101, 123, 161; flight 320 9–10
American Airways 101
An American in Paris 79
"American Pie" (song) 10
AND *see* aircraft nose down
Anderson, Eddie "Rochester" 79
Anderson, Eric 48, 49, 50
The Andy Griffith Show 18
ANU *see* aircraft nose up
"Anvil of the Gods" 160
Aono Domon (Blue Tunnel) 170
APC *see* Aircraft-Pilot Coupling
Arkansas 9–10, 122, 126, 188
Arlington, NE 188
Armstrong, Louis 79
Army Air Forces 71
Army Air Service 56–57, 73
Arnaz, Desi 80
ARTCC 27, 67
artificial horizon indicator 50
ASM *see* available seat miles
Assembly of the French Union 77
Associated Press (AP) 138
Astrojet 123
Atheneum Reunion 71

Atlanta, GA 52, 56, 64–65, 67, 82, 97
Atlantic Ocean 11, 12, 15, 18
attitude indicator 41, 95; *see also* HZ-75
Australia 163
autogiro/Autogyro 56
automatic trim coupler 89
autopilot 10–12, 14, 16–17, 63, 88–89, 129, 180
available seat miles (ASM) 116–117
Aviation Corporation of Delaware (AVCO) 100
Aviation Medical Acceleration Laboratory (AMAL) 49, 51, 91–93

BAC-Aerospatiale Concorde 103
BAC One-11 *see* British Aircraft Corporation One-11 (1-11) jet
The Bahamas 104
Bain, Donald 107
Bainbridge, PA 96
Baker, Dick (CAB) 160–162
Baker, Dick (mayor of Falls City) 156
balance bay icing 37–38
Baldwin, Henry 42
Ball, Lucille 80
Baltimore, MD 104
Barksdale Air Force Base 61–62
barometric pressure 173
Bataan 70, 79–80
Bataan, Philippines 72, 79
Bataan Death March 79
Batelle Institute 37–38
Bauman, Dennis 162

213

Index

Bayard, NE 186
Beard, Charles F. (Chuck) 102–103, 113
The Beatles 6
Beatrice *Daily Sun* 159
Becker, Elmer 162
Becker, Wayne 146
Beechcraft Bonanza 10
Belgium 77
Bellevue, NE 186
Berlin, Germany 73–75, 124
Bermuda 58
Berwyn, IL 43
Birth of a Jet 54
Blackman, Honor 110
Blair, NE 188; First Methodist Church of 188
Blank, Connie Rae 24, 42
Bloomington, MN 7, 8
Blue Tunnel *see* Aono Domon
BOAC *see* British Overseas Airways Corporation
"Bockscar" 170
Boeing B-17 Flying Fortress 73–75, 124–125
Boeing B-47 9
Boeing Commercial Aircraft Co. 2, 16, 19, 37, 38, 40, 41, 45, 49, 53–54; *see also* 707: *Year One*
Boeing 707 9–12, 15–16, 18–19, 52, 62–63, 101, 109, 114, 116, 163; -121 11, 18; -200 101; -320 C 114; *see also* 707: *Year One*
Boeing 720 58
Boeing 720-B 21, 23–24, 26, 28–29, 31, 34–35, 37–38, 41–42, 47–48, 51
Boeing 727 58, 114, 119, 180
Boeing 727 QC 114
Boeing 747 105
Boeing 747-8 3
Boeing 787 Dreamliner 54
Boeing Stratoliner 10
Boeing 2707 105–106
Boise, ID 104
Bonanza (TV show) 18
Bond, James 110
Bosted, Larry Joseph, Private 186
Boston, MA 58, 78, 101, 104, 114
Bowler, Jack Kerry 96
Bowman, Lee 80
Boy Scouts 75
Brandenburger Tor, Germany 73
Braniff, Paul 100–102
Braniff, Paul R., Inc. 100–101

Braniff, Thomas (Tom) 100, 102
Braniff International Airways 1–2, 8, 45, 58, 83, 99–104, 106–119, 121–125, 128–131, 133, 135–142, 144, 146, 148–149, 151–152, 154, 156–161, 163, 166, 168–170, 172, 174–178, 180, 183–187, 189–192; Conquistador Service, 102; flight 146 156; flight 226 156; flight 250 1–2, 45, 83, 100, 115, 119–120, 122, 124–125, 129–132, 134–138, 140, 142, 144, 148–149, 152, 156–160, 163, 166–170, 172–174, 177–178, 181–188, 191–192; flight 251 122; flight 255 133, 167, 175; flight 352 184–185
Brennan, Dr. L.V. 138
Brennan, Richard 138
Brightwell, Christine 162
Brisbane, Ginger 2, 129–130, 137, 156, 167, 186
Brisson, Gabrielle M. 96
Brisson, Marie M. 96
Brisson, Peter V. 96
Brisson, Timoteo V. 96
Bristol Aeroplane Company 118
British Aircraft Corporation 1; One–11 (1–11) Jet 113, 116–123, 128–130, 132–133, 135, 139, 145, 147, 156–157, 159, 161, 168, 174–175, 179–181, 184, 186, 188, 193
British Airways 106
British Overseas Airways Corporation 119
British United Airways 121
Broadfoot, Andrew Dewitt 186
Broadway 79–80
Broman, Signe 42
Brown, Kenny 162
Brownsville, TX 101
Brunswick, Germany 74
Buchenwald (Concentration Camp) 76, 78
Buenos Aires, Argentina 102, 112
Buffalo, NY 56
Bunch, Carl 10
Burgess, B.F., Jr. 49
Burrwood, LA 67
Butterfield 8 18
Byczynski, Carl A. 96
Byczynski, Joan Marie 96
Byczynski, Sophie T. 96
Byers, Horace 172, 174

CAB *see* Civil Aeronautics Board
Cabin in the Sky 70, 79–80
Cain, Joseph E. 42
Calcutta, India 119
California 17, 53–54, 75, 88
Callan, Clair 169
Camblin, Lanthan (L.D.) 136, 144
Cambridge, UK 73
Camp Beauregard, LA 71
Camp Robinson, AR 126
Canada 58
Canadian Pacific Airlines 54
Cape Town, South Africa 3
Capital Airlines 118
CAR *see* Civil Air Regulations
Caracas, Venezuela 43
The Caretakers 18
Caron, Leslie 70, 78
Cary, IL 43
Case Red (Fall Rot) 76
Case White (Fall Weiss) 76
Case Yellow (Fall Gelb) 76
Castle, Haskell Louiese Sherman 96
Castle, Walter Francis 96
Central America 52
Chab, Bob 146
Chalmette, LA 97
Chambers, Randall M. 49
Chamblin, Nancy 186, 188
Chamblin, Susan 186, 188
Chaney-Hodgen-Sharrar Funeral Home 138
Charles Butler Associates 123
Charlotte, NC 2, 97, 178
Cherbourg, France 73
Chicago, IL 21, 22, 25–26, 42–43, 47–49, 51, 58, 97, 99, 101–102, 104, 114, 131, 133, 134, 149, 164, 165, 167, 172, 175, 186; Air Route Traffic Control (ARTC) 133; O'Hare International Airport 47, 49; Wicker Park 164
Chicago and Southern Airlines 99
Chicago Daily News 164
Chicago Sun-Times 164
Chicago Times 164
Chown, Harriet Leehahn 96
Chown, Roger Murray 96
Christianson, Wilbur L. 42
Christianson, Mrs. Wilbur L. 42
Cicero, IL 43
Civil Aeronautics Act of 1938 99
Civil Aeronautics Board

(CAB) 10–13, 16, 18, 22, 24–25, 28–32, 34–36, 38, 40–42, 44–45, 59–60, 62–63, 65–66, 69–71, 81–87, 89–96, 104, 114, 136–137, 145, 147, 156, 160, 161–168, 172–173, 177–178, 184
Civil Air Regulations (CAR) 93
Civil Aviation Regulations 17
Civil War 75
Clark, Harry 9
Clear Lake, IA 10; Surf Ballroom 10
Clifton, J.N. 149, 163
Coates, Eddie 65
Cockpit Voice Recorder (CVR) 31–33, 83, 161, 163–164, 166, 168, 182
Coffee, Tea or Me? 107
Colbert, Jim 138
cold soaking 37–38
Collar, Charles S. 31
Collier Trophy 56
Collins, Larry 78
Collins, Vernon Arthur 96
Collins Company 93; 105 Approach Horizon 89
Colorado Springs, CO 122
Columbia Masterworks 80
Columbus, MS 72
Comer, Alvin 146
Comité de Libération (Committee of Liberation) 76
Concorde 106, 112, 121, 172
Coney Island, NY 19; Cyclone 19
Connery, Sean 110
Constituent Assembly (France) 77
Continental Airlines 99, 102, 107, 178; flight 426 178
Convair 45, 113, 116, 123, 175
Convair 580 175
Convair 880 123
Convair-Liner 113
Cook, Beulah 96
Cook, Joe James 96
Cook, Noel 96
Copp, Martin R. 20
Coppola, Francis Ford 78
Council Bluffs, IA 186
Coward, Noel 107
Cox, Danny Ray 186
Cox, Jere 152
CPS *see* cycles per second
Craig, MO 163
Crawford, Joan 18
Croix de Guerre (War Cross of 1914–1918) 76

Croix de la Libération 76
Crosbie, Richard J. 49
Cuba 43
Curtiss, Glenn 56
Curtiss C-46 67
CVR *see* Cockpit Voice Recorder
cycles per second (CPS) 90

D-Day 74
Dade County, FL 30
Daley, Richard J. 165
Dallas, TX 2, 58, 100–104, 106, 115, 122, 152, 160, 170, 178, 185
Dallas–Fort Worth International Airport 2
Dallas Love Field 102, 115
Dalton, GA 96
Damon, TX 61
Dawson, TX 184–185
Day, John 75
Decatur, TX 186
de Gaulle, Charles 76
de Havilland 119
de Havilland Comet 119–121, 168
de la Cierva, Juan 56
Delta Air Lines 2, 53–55, 99–100, 123, 156, 178; flight 191 2, 178
Denies, Ronald L. 186
Denmark 18
Denver, CO 101, 104, 178
Department of the Aisne (France) 77
Department of Transportation 117
de Saint-Exupéry, Antoine 5, 128
design gust 181–183
Des Moines, IA 78, 122, 167, 176, 186
"Detailed Investigation of Mesometeorological Conditions of the Squall Line of August 6–7, 1966, Which Crossed the Air Route Between Kansas City, Missouri and Omaha, Nebraska" 172–173
Detroit, MI 96
Diaz, Ramon 43
Diesel Eight 52–53, 58
Dillon, Lt. Cdr. James 29
DiMucci, Dion 10
The Disorderly Orderly 18
Distance Measuring Equipment (DME) 27
Dr. Kildare 18

Donovan, Sgt. Sherman J., Jr. 96
Dorr, Russell 138
Dorr-Philpot Funeral Home 138
Douglas, Don 54
Douglas Aircraft 10, 23, 45, 52–55, 58–59, 63–64, 75 86–88, 93, 96, 101–102, 119, 180
Douglas DC-3 11, 23, 65
Douglas DC-4 11, 23, 53
Douglas DC-6 11, 23, 52, 65, 108, 116
Douglas DC-7 10–11, 23, 46, 52, 58, 65, 101; DC-7C 52, 116
Douglas DC-8 44, 52–55, 58–65, 83–87, 90, 93–94; *see also* Diesel Eights; DC-8-11 55; DC-8-12 55; DC-8-21 55, 59, 63, 96; DC-8-42 53; DC-8-61 55; DC-9 119, 180; MD-83 88
downdraft(s) 37, 169, 176–177
Dubuque, IA 3, 46
Duck Creek, WI 47
Duerkson, Jean 186
Duescher, Lynden E. 47, 49, 50
Dusseldorf, Germany 96
Dutch roll 23
Dyer, Ava 186

Earl, Alan Lloyd 97
East Coast airline 57
East Prussia 73
Eastern Air Lines 2, 34, 44, 52, 55, 57–59, 62–67, 69–71, 74–78, 80–84, 86, 88–91, 92, 94–97, 99–100, 122–123, 164, 169, 178; flight 66 2; flight 300 ; flight 301 44, 49, 52, 58–63, 65, 84, 88–91, 94–95; flight 304 49–50, 58, 64–67, 77–78, 80–86, 88, 91–92, 94–96, 122–123; flight 305 64
Eastern Air Transport 57
Eastern Airlines 99
Eastman School of Music *see* University of Rochester, NY
École des Sciences Politiques 75
Economic and Social Council (France) 77
Eden Prairie, MN 188, 192
Educational Symposium 71
Edwards Air Force Base 53–54
Eighth Air Force 73, 126
Electra *see* Lockheed L-188

elevator control tabs 87
elevators 87
Elkton, MD 163
Elmhurst, IL 48
Elroy, M.C. 61
The End of the Plain Plane 100, 107–108, 110, 112–113, 115, 123
Engebretson, Wendy F. 24, 42
England 18, 73–74, 76, 122
English Electric Aviation 118
Enloe, George A. 43
Epps, Private Wesley 80
equivalent air speed (EAS) 182
Erding, Germany 73
Eschback, Donald 186
Eskelinen, Kenneth 186
Espiritu, Roque 80
Europe 72
European Theater of Operations 74
Ewert, Myrna E. 24, 42

F-Scale *see* Fujita Scale of Tornado Intensity
F-WTSA 106
FAA *see* Federal Aviation Administration
Fairchild 32, 35, 83; *see also* Flight Data Recorder
Fall Gelb (Case Yellow) 76
Fall Rot (Case Red) 76
Fall Weiss (Case White) 76
Falls Church, VA 96–97
Falls City, NE 5, 6, 8, 83, 100, 102, 104, 106, 108, 110, 112, 114–116, 118, 120–122, 124–126, 130, 132, 134, 136, 138–156, 158–164, 166, 168–170, 172–174, 176, 178, 180, 182, 184–186, 188–191
Falls City Breezy Hill Drive-in 162
Falls City Community Hospital 149
Falls City Country Club 6, 125
Falls City *Journal* ix, 124–125, 134, 136, 138–155, 158, 160–162, 168–169, 188, 190
Falls City Municipal Airport 143
Falls City Prichard Auditorium 6, 140, 150, 152, 158, 182, 188, 190
Falls City Saints Peter and Paul Catholic Church 138
Falls City Volunteer Firemen's Auxiliary 152
Falter, John Phillip 161
Family Weekly 12

Fargo, ND 156, 186
Farmer's Short Course 71
Fastback Jet 113, 115, 117, 119, 121, 123
"Fat Man" 170
Father Knows Best 18
Father of the Bride 79
Faust 79
Fayette, IA 78
Fayette County *Leader* 78
FDR *see* Flight Data Recorder
Federal Aviation Administration (FAA) 17, 22–24, 34, 42, 49–50, 63, 81, 106, 129, 161–165, 183
Federal Aviation Regulations (FARS) 17; Part 41.62 17; *see also* Civil Aviation Regulations
Federal Bureau of Investigation (FBI) 30, 32, 138, 149, 151, 153–154, 161
Feller, Robert J. 23, 26–27, 36, 42
Ferrero, Donald 186
Fife, Barney 18
Finn, Huck 19
509th Squadron 73
Flamingo, FL 29
Flight Data Recorder (FDR) 26, 31, 32, 33, 35, 36, 37, 40, 83, 161, 163, 165–166, 175
Flight International magazine 123
Flight magazine 52
Flight Traffic Dispatchers Association 161
Florence, Italy 108
Florida 21, 26, 58; Everglades 28, 29, 31, 34, 35, 42, 47, 51; Highway Patrol 30
Flowers, Grover Wesley "Dusty" 96
Flushing Hospital 10
Flying Colors 106, 185
Flying magazine 54
Focke-Wulf 74
Forbes Air Force Base, KS 146
Forces Françaises de l'Intérieur *see* French Forces of the Interior
Forensic Banquet 71
Forest City, MO 162
Fort Lauderdale, FL 22, 32
Fort Myers, FL 25
Fort Polk, LA 131
Fort Riley, KS 156
Fort Smith, AR 103, 106, 122, 186, 188
Fort Worth, TX 2, 101, 186

Fortescue, MO 162
Foster, Leslie David, Jr. 186
Foster, Sgt. Lyle 146
4-H Club 6, 140, 150–151; Fair 6, 150–151
Fourth Republic *see* France
France 17, 56, 73–74, 76–77, 80
Frandsen, Ray 146, 149
Freedom of Information Act 137
Fremont, NE 186
French, Mel 59–62, 84
French Army 75
French Economic and Social Council 77
French Forces of the Interior 76
French Légion d'Honneur 17
French National Assembly 75; Committee of France Overseas 77; Foreign Affairs committee 77
French National Senate 75
French resistance 70, 78
French Union 77; Assembly 77
Fresnes, France 76
Friesen, Allen R. 24, 42
Frontier Airlines 175; flight 564 175
Fujita, Dr. Tetsuya "Ted" 1, 2, 45, 168–178, 181–183
Fujita, Tomojiro 170
Fujita, Yoshie 170
Fujita Scale of Tornado Intensity (F-Scale) 2, 178

G forces 14, 33, 36, 40, 49, 50
Galler, Dan 43
Galore, Pussy 110
Gander, Newfoundland 11, 12, 15, 16, 18
Gandt, Robert 105
Garden of Memories Cemetery, LA 96
Garland, Judy 79
Geheime Staatspolizei *see* Gestapo
The Gene Autry Show 7, 188
The Gene Krupa Story 18
The General 79
General Dynamics 119
General Motors 57
German (language) 80
Germany, Federal Republic of 70, 73–74, 76–78, 80
Gershwin, George 17; *Concerto in F* 17
Gestapo 76, 78

Index   217

Gifford, Martin 136, 128, 142
Gilbertson, Patricia 186
Girard, Alexander 108–110, 115, 117, 125, 130, 141, 147
Glen Flora, WI 186
Gminski, Anne M. 97
Gminski, Louise A. 97
Gminski, Stephen M. 97
Go Team (CAB/NTSB) 160–161
Goethe *see* von Goethe, Johann Wolfgang
Golden Falcon Jets 52, 58, 123
Golden Fleet 55
Goldfinger 110
Gomer Pyle, USMC 18
Gonzalez, TX 186
Goodwin, Ted 43
Goodwin, Mrs. Ted 43
Graeber, Lyman Monroe 186
Grand Rapids, MI 43
Grantland Theater 75
Great Depression 153, 164
The Great Silver Fleet 52, 58
Great Weaver Hotel, Falls City 161
Greatamerica Corporation 102–103
Green, Rev. 79
Greene, Stewart 108
Greenwood, MS 72
Griffin, GA 97
Ground Zero (Nagasaki) 171
Grumman Mallard 102
Guatemala City, Guatemala 156, 186
Gulf of Mexico 64, 70, 82
Gulfstream 161
Gummers, Mrs. G. 186, 188
gust force 182
gust intensity 181
gust load 54
gust velocities 180

H-37 Helicopter 31
Hailey, Arthur 70
Halaby, Najeeb 45, 168
Hamm, Mary Kay 186, 188
Hamm, Susan 186, 188
Hampton, GA 96
Hapeville, GA 96
Harahan, LA 96
Havana, Cuba 102
Heathrow Airport, London 11
Heil, John C. 43
Heimerdinger, A. G. 54–55
Hendricks, Sharon 2, 129–130 137, 156, 167, 186
Hersh, Tim ix
Hiawatha, KS 162

Hickam Field, HI 72
Hillcrest Baptist Church, Omaha 159
Hiller Aviation Museum 105
Hilliker, Bob 191
Hilliker, Daniel (Dan) 8, 188, 190–193
Hilliker, David 8, 191
Hilliker, James A. (Jim) 2, 8, 129, 131–132, 137, 156, 166–168, 183–184, 186–193
Hilliker, Patricia 8, 188–190
Hilliker, Sandy 191
Hilliker, Spike 191
Hilliker, Timothy "Tim" 7, 187, 189, 192
Hilliker Prestegard, Kathleen (Kathy) 8, 191–192, 202
Hinkle, Alton 162
Hinkle, Wendell 162
Hiroshima, Japan 170–172
Hitchcock, Lloyd, Jr. 49
Hitler, Adolf 71, 73, 78
Hoban, Dr. Bob 153
Hobby, William P. Airport 60
Hodgens, Bill 138
Hollerich, Jack P. 43
Holly, Buddy 10
Hollywood 12, 17, 21, 70, 78–80, 105, 121
Hollywood Bowl 79
Home Safety Council 3
Honea, Walter, Cpl. 146
Hoover, Herbert 57
Hoover, J. Edgar 149
horizontal stabilizer 87–89; jackscrew 88–89
Horne, Lena 79–80
Horse Play Days 6,
Horton, KS 162
Hoskins, Charles 75
Hotel Stephenson, Falls City 161; Gold Room 161
Houston, TX 58, 61, 90, 101, 185–186, 188; Intercontinental Airport 61; International/Hobby Airport 60
Howard, Charles E. 186
Hudson, Russell E. 186
Hughes, Dan 152
Hullman, Mrs. Lawrence 140
Humboldt, NE 138
humidity 54, 173
Hunting Aircraft 118
Huntingdon Station, NY 60
Hurlburt, Claire 134–136, 138, 142, 190
Hurn, UK 121
Hussein, King of Jordan 45
HZ-75 Attitude Indicator 41

IBM 37; 7094 mainframe computer 175
Idlewild Airport, NY 11, 58–59, 63
Idol, Harry 64–66, 96
IFR *see* Instrument Flight Rules
Illinois House of Representatives 164
Independent Scheduled Air Transport Operators' Association 101
Indian Ocean 119
Indianapolis, IN 42
indicated air speed (IAS) 33
Ingram, Rex 79
Inland Airlines 99
The Inspector General 17
International Center for Air Transportation 116
International Civil Aeronautics Organization (ICAO) 44, 45
International Council of Women 68, 77
International Music Festival 80
Internet 80
Interstate Host, Inc. 81; "International Room" 81
Instrument Flight Rules (IFR) 66
Iowa 78
Irwin, W.R. 43
Is Paris Burning? 70, 78
Italy 80

J-37 air route 67
Jackson, Little Joe 79
Jackson Memorial Hospital, FL 30
jackscrew *see* horizontal stabilizer
Jacobson, Patricia 186
Jamaica 104
James, Mark 162
Japan 164, 170–171
Japanese Imperial Army 79
Japanese Imperial Navy 171, 175
Jennings, Waylon 10
Jensen, Tove Emma 96
Jergenson, Larry Dean 97
The Jet Age 18, 19, 20, 45, 46, 52, 57, 58
Jet Clipper Washington 13
Job, Macarthur 125
John Henry (musical)
John Marshall Law School 164
Johnson, Frederick H. 97
Johnson, Lyndon B. 165

218 Index

Johnson, Mabert H. 97
Johnson, William O. 186
Johnsville, PA 36, 49
Jones, Rex, M.Sgt. 146
Jordan, Cheryl Lyn 186–187
Jordan, Jeff 187
Joy, Bob 162
JT3C engines 55
JT4A engines 55; series -9 59, 83

Kane, Patrick Kelly 97
Kansas City, MO 101, 116, 122, 129–133, 137, 143–144, 148–149, 156, 159, 166, 170, 173, 176, 179, 187; Air Route Traffic Control (ARTC) 132; Metropolitan Airport 130–131; Zoo 132
Katyn Forest 165
Keaton, Buster 21
Kelinson, M.D. 43
Kelinson, Mrs. M.D. 43
Kelly, Gene 9–18, 79
Kelly, Kerry 11
Kennedy, John F. 58, 105, 165
Kennedy International Airport, NY 58, 65
Kenner, LA 2, 128
Kennilworth, IL 43
Kersey, John 152
Keys, Clement 57
Kirschenheuter, Fred 97
Kitakyushu City, Japan 170
Koenig, Frank 144
Koenig, Henry 144
Kokura, Japan
Kowtaliw, Bohdan 186
Kubrick, Stanley 105
Kuhlman, David, Lt. 146
Kuhr, Melinda Sue 188
Kuhr, Mitchell 132, 186–188
Kuhr, Robert 187–188
Kuhr, Ruth L. 132, 186–188
Kurnitz, Harry 10, 11, 12, 13, 17
Kuttler, Bill 135
Kuttler, Marcelle 135
Kuwait 96
Kyushu, Japan 172
Kyushu Institute of Technology 170, 172

LaBelle, FL 24
La Fonda del Sol 108–109
LaGlacerie, France 73
LaGuardia Airport, NY 9
Laird, John 11, 14
Lake Pontchartrain, LA 64–65, 68, 69, 78, 82, 86, 89, 94–95

Lake Success, NY 77
Lamb, William L. 70, 84
Lancaster, WI 75
Langley Research Center 20, 181; see also National Air and Space Administration (NASA)
Lapierre, Dominique 78
Larsen, Agnew 56
La Salle, IL 43
Lautman, Les 2
Lawrence, Harding L. 102–110, 112–114, 117, 119, 123, 189
Lear artificial horizon indicator 50; see also artificial horizon indicator
Lebodow, Mrs. Fanny 43
Lefaucheux, Marie-Hélène (née Postel-Vinay) 69, 75–78, 95, 97
Lefaucheux, Pierre 75
Légion d'Honneur 76
Leipzig, Germany 74
Le Mans (company) 76
Lewis, Jerry 18
Lieurance, Newton 45
Life magazine 45–46, 168
Lincoln, NE 149, 169, 188
Lincolnwood, IL 43
Lindseth, Bob 8, 189, 191
Little Banana Patch, FL 28, 29
Little Rock, AR 186
Little Rock Air Force Base 9
load factor(s) 106, 116–118
load management 58
Lockheed 9–10, 23, 45, 52, 65, 101, 103, 105, 119, 123, 156, 168, 185
Lockheed L-10 Electra 101
Lockheed L-188 Jet-Prop Electra 9–10, 23, 45, 65, 101, 123, 156, 158, 185
Lockheed L-1011 Tri-Star 123
Lockheed L-1049 Constellation 52
Lockheed L-2000 105
Lockheed Vega 101
London, UK 11–12, 18, 73, 106, 160; Heathrow Airport 11
London Airports 11; see also Heathrow Airport
Long Beach, CA 54, 59
longitudinal control 85–86
longitudinal trimming 88
Los Angeles, CA 9–10, 79, 97, 101, 106, 114
Los Angeles International Airport 9
Louisiana 66
Lubbock, TX 122

Ludington Air Lines 57
Luxury Liner 123
Lynch, Waldo 11, 13, 14, 15

Mach 53; meter 14; warning bell 14
Mackey, Dispatcher 11, 14
Macon, GA 25
Madison, WI 72
Madison Capital-Times 71
Magic Washer soap 164
Magruder, Bill 53
Mahoney, Ed 142
Mailwing 56
Manhattan, NY 60
Manilla, Philippines 79
Marmet, Cheryl 162
Martin 2-0-2 65
Martin 2-0-4 65
Martin 4-0-4 75
Marx, Groucho 117
Maryville, MO 173, 175–176
M*A*S*H 18
Massachusetts Institute of Technology (MIT) 116
Maxwell Air Force Base, AL 72
May and Timm funeral home 138
Mayer, Adolph 186, 188
McCabe, Rev. John 138
McClement, Fred 160
McCloud, Glen 149, 163
McConnell, Eugene P. 186
McKim, Ray, Jr. 146
McKimmey, Ray, Spec. 146
McKnight, Duane 138
McLean, Don 10
McQueen, Alexander 117
Measuring Worth (web site) 48
Médaille de la Résistance 76
Mediterranean Sea 119
Meet Me in St. Louis 79
Meiji College 170–171
MEL see "Minimum Equipment List"
Melahn, Arnold 43
Melahn, Mrs. A. 43
Meltzer, Louis G. 97
Meltzer, Natalie S. 97
Merseburg, Germany 74
Mesometeorology 2, 174
Metairie, LA 96
Mexico 52, 58, 64–65, 67, 102, 110–111, 114
Mexico City, Mexico 44, 52, 58–65, 67, 77
MGM 79–80
Miami, FL 19, 22–31, 52, 56, 104, 106, 114

Index

Miami International Airport, 19, 21, 26, 28, 30, 35
Miami VORTAC 24, 28
Micosukee Seminole Indian Settlement 30
Mid-Continent Airlines 99, 114, 122, 129, 192
Military Airlift Command (MATS) 114
Miller, Austin "Dusty" 1
Miller, Herman 108–110, 125, 130
Mills, Opal 186
Mineo, Sal 18
"Minimum Equipment List" (MEL) 84
Minneapolis, MN 7–8, 23–25, 42, 58, 122, 129, 132, 156, 159, 186–187; International Airport, 23; Roosevelt High School 23
Minneapolis *Star-Tribune* 8
Minnesota Vikings 187
Minnelli, Liza 79
Minnelli, Vincente 79–80
Mississippi River 19
Missouri River 127, 163, 173
Mr. Tornado 169–178
Mitchell, Thomas 79
Mohawk Airlines 121, 161
Montgomery, AL 72
Montreal, Canada 58, 97
Moorhead, MN 10
Morgan, Jean M. 97
Morgan, Joseph Warner 97
Morgue, Dade County, FL 30
Morrison, Frank, Gov. 149
Moscow, USSR 71
Munich, Germany 73
Municipal Council of Paris 77
Murphy, George 79
Murphy, SSgt. Glenn 146
Murphy, William 186

N1553 115, 129–132, 139, 145, 159, 163, 168, 172, 177, 180, 182–183, 186, 190, 193
N712PA 9, 17–18
N724US 21, 26, 31, 34, 35, 37, 42
N8603 59–60, 63
N8604 59
N8607 64–65, 83–84, 89–91
NAACP *see* National Association for the Advancement of Colored People
Nagasaki, Japan 170–172
Nance, John J. 100, 103
Naples, Italy 80
Nashville, TN 72

Nassif, Clarice Mary 97
National Air and Space Administration (NASA) 20, 34, 40, 181; "Conference on Some Problems Related to Aircraft Operations" 20; wind tunnel tests 40
National Airlines 97, 99
National Association for the Advancement of Colored People 80
National Basketball Association 105
National Car Rental 102
National Guard *see* United States National Guard
National Severe Storms Project (NSSP) 34–36; *see also* United States Air Force
National Socialism 73
National Socialist German Workers Party *see* Nationalsozialistische Deutsche Arbeiterpartei
National Transportation Safety Board 36, 88, 95, 125, 128–129, 136–137, 184–185
Nationalsozialistische Deutsche Arbeiterpartei (NSDAP) *see* National Socialist German Workers Party
Naval Medical Research Institute (NMRI) 36; *see also* United States Navy
Nebraska Highway Patrol 136, 138, 142, 145, 161
The New Deal 57
New Mexico 108
New Orleans, LA 52, 58, 64–67, 80–83, 89, 94, 96–97, 122, 128, 130, 178, 189; ARTCC 67; Center 67; Lakefront Airport 83; Moisant (Louis Armstrong) International Airport 65–67, 80–81, 83, 89, 96, 122, 128, 130; VORTAC 67
New York Airlines 57
New York City, NY 2, 9–11, 13, 17–19, 48, 52, 56–60, 63, 65–66, 70, 77–79, 84, 96, 103–104, 107–110, 114, 123, 134, 159, 178
New York City airports 9, 11, 52; Idlewild 52, 58–59; Kennedy 58, 65; LaGuardia
New York Times 134, 159
Newby, Dr. Grant 72, 74
Newby, Grant R. 60–62, 64–75, 84, 88, 90, 94–95, 96

Newby, Mrs. 72, 74
Newark, NJ 104
Newgulf, TX 61
Newport, TX 59, 63
Nienburg, Germany 74
Night Flight 128
Niles, IL 43
Nimsch, Fred 43
Nolan, Lloyd 79–80
Noll, Jun, Sgt. 146
Norman, Barbara Delane 96
Normandy, France 73
North Africa 71
North American Aviation 57
North Little Rock, AR 186
Northeast Airlines 99, 118
Northwest Airlines/Northwest Orient Airlines 19, 21, 28, 34, 44, 46, 38, 42, 48, 58, 81, 83, 99, 164; Batelle joint study 37–38 (*see also* Batelle Institute); flight operations manual 25; flight 700 21–25; flight 705 19–46, 48–49 62, 82–83, 89; Regal Imperial Service 21–23; Ship 724 22, 31 (*see also* N724US); turbulence penetration procedure 38
Northwestern University 164
NTSB *see* National Transportation Safety Board
Nutzman, John 135

Oak Lawn, IL 48
Odyssey: A Daring Transatlantic Journey 18
Offutt Air Force Base, NE 186
Okinawa 75
Oklahoma 100–102, 105, 122, 187
Oklahoma City, OK 101–102, 105
Oklahoma City Thunder 105
"Ol' Man River" 80
Oliver, Susan 10, 11, 12, 13, 17–18
Olson, Fred, III 43
Olson, Joan 43
Omaha, NE 43, 116, 122, 124, 130–133, 136–137, 156, 159, 166, 170, 173, 175, 186–191; Eppley Field 187–188
One-11 (1–11) jet *see* British Aircraft Corporation
134th Infantry 126
O'Neill, NE 47–48, 51
Opa-Locka, FL 31; Airport 31
Opéra-Comique 17
Oppenheimer, Peer J. 12

Orszule, Jerilyn 43
Orszule, Walter 43
Orszule, Walter, Jr. 43
Orszule, Mrs. Walter 43
Oschersleben, Germany 74
Overland Park, KS 186
Owosso, MI 56
Owosso Sugar Company 56

Pacific Ocean 88
Pan American World Airways 10, 11, 16–18, 20, 105, 121, 123, 164; Clippers 123; flight 115 9–11, 13, 16–18; flight 214 163–164; flight 759 2, 178
Panama Canal Zone 102
Parade magazine 12
Paris, France 10–12, 75–77, 97, 103, 106, 121; Municipal Council 77
Park Ridge, IL 43
Parrotsville, TN 162
Pas de Deux 17
Paul, John H. 186
Pauly, Donald G. 2, 129–133, 137–138, 159, 166–168, 182–184, 186, 191
Pauly, Donald G., Jr. 129
Pauly, Earl 129
Pauly, Linda 129
Pauly, Noel 129
Pawlowski, Carolyn B. 97
Pawlowski, Harry J., Jr. 97
Pawnee (City), NE 167, 183
Pearl Harbor, HI 72
Pennsylvania-Central Airlines 99
Pensacola, FL 66
Peterborough, UK 73
Peters, Samuel 11, 13, 14
Peterson, Roger 10
Peyton Place 18
Philadelphia, PA 61, 78, 104
The Philippines 79
Phony War 76
Pilot-Induced Oscillation (PIO) 51, 91–93
Pilot Report 26, 28; *see also* PIREP
The Pink Panther 17
PIO *see* Pilot-Induced Oscillation
Pioneer Players Play 71
Piper, Louis H. 100
PIREP 26, 28
Pitcairn, Harold 55–57
Pitcairn Aircraft Company 56; Mailwing 56;
Pitcairn Aviation 57

Pitcairn-Cierva Autogyro Company of America 56
pitch trim actuator 88
pitch trim compensator (PTC) 66, 84, 87–88, 90, 93–95
pitch trim potentiometer (PTP) 17
Pittsburgh Plate Glass (PPG) 55
Platteville, WI 71-; High School 71; State Teachers' College 71
Plautz, Ethel Adams 97
Plautz, Louis 97
Playhouse 90 18
Pokorny, Ed 146
Poland 76
Polebrook, UK 73–74
Polish Museum of America 165
Ponca City, OK 101
Popkess, Robert 149
Popkess, Roger 149
Popular Mechanics 26
Portland, OR 43, 96, 104
Post, Troy 102
Postal Service *see* United States Postal Service
Postel-Vinay, André 76
Postel-Vinay, Roger 76
Pratt and Whitney engines 22, 84
precipitation 173–176
press pool reports 47, 64
Prichard Auditorium *see* Falls City Prichard Auditorium
Proctor, Bob 115
Proctor, Jon 115
PTC *see* pitch trim compensator
PTP *see* pitch trim potentiometer
Pucci, Emilio 108–111, 130, 137, 188
Pucinski, Lidia 164
Pucinski, Roman C. 163–166
Puerto Rico 52, 58, 104
Purvis, Dale 136

radar 20–22, 26–29, 33, 42, 59, 66–68, 82–83, 109, 132, 173, 175–176, 181–184
Ragland, Stuart, Jr. 49
Ramsey, NJ 97
Rand, A.B. 43
Rawhide 18
Reavis-Macomber funeral home 138
Reed, Lt. Merlin 125
Renault 76

revenue passenger miles (RPM) 116
Rever, Christine 43
"A Review of Atmospheric Turbulence and Its Significance to Jet-Transport Operations" 20
Rhea, Gilmore 43
Rhea, Mrs. Gilmore 43
Ricardo, Ricky 80
Richardson, J.P. "The Big Bopper" 10
Richardson County, NE 136, 142–143, 152, 155, 173
Richardson County Civil Defense 144, 149
Richmond, J.O. 144, 149
Rickenbacker, Eddie 52, 55, 57, 58, 75
Ringoes, NJ 65, 96
Rio de Janeiro, Brazil 106
Robert, Jules Charles 97
Robertson, Garrett George 186
Robeson, Paul 70
Rochlitz, Ervin A. 48
Rockford, IL 43
Roettger, Grace Rhodes 186
Rogers, Mrs. Elmer 140
Rolls-Royce 119, 132, 145, 154, 180; Spey engines 119, 129–130, 145, 154
Rome, Italy 119
Romig, Eugene A. 74
Rookery Creek, FL 29
Roosevelt, Franklin 57
Rose Pine, LA 97
Rosemont, MN 23
Route 66 18
Rowland, Basil 78
Royal Crown DC-7 123
RPM *see* Revenue Passenger Miles
Rulo, NE 162
Russians 6
Rygaard, Ole F. 97

Saarbrucken, Germany 74
Sailors, Blaine 136
Saint-Dizier, France 76
St. Joseph, MO 161, 176; Rosecrans Airport 161
St. Louis, MO 15, 58, 79, 100–101; Civic Opera Company
Saint-Quentin-des-Prés, France 78
Saints Peter and Paul Catholic Church, Falls City 138
Saipan 171
Salt Lake City, UT 104, 155, 162

Index 221

San Antonio, TX 101
San Carlos, CA 105
San Francisco, CA 19, 47–48; International Airport 47
San Juan, PR 58
Sandell, Mary S. 24, 42
Sarau, Germany 74
Saturday Evening Post 161
Sauk Village, IL 186
Scandinavia 77
Scharry (or Sharry), Sgt. E. 97
Schary, Dore 80
Schawang, Antone "Tony" 2, 7, 127, 134–136, 142, 153, 159, 179, 182, 187, 189–191
Schawang, Barney 135
Schawang, Elaine 189–191
Schawang, Kenny 189–193
Schawang, Paul 135
Schawang, Roy "Bud" 135
Schawang, Vernell 2, 7, 138, 140, 149, 189, 191
Schawang family 136, 162
Schawang farm/farmhouse 134–136, 138, 140, 142, 149–153, 155, 159, 179, 182, 187, 189–191, 193
Schepman, Henry F. 152
Schkeuditz, Germany 74
Schmidt, Carl F., M.D. 49
Schock, George William (Bill) 124–127, 134, 136–140, 142–144, 149–155, 159–162, 164, 185, 187–191
Schock, Scott 124
Schreiner, David 75
Schutzstaffel (SS) 76
Schweinfurt, Germany 74
Schwendener, Susan 43
SD 76; *see also* Sicherheitsdienst des Reichsführers-SS
Seattle, WA 16, 22, 43, 104–105, 114
Seattle SuperSonics 105
segregation 17
Sellers, Peter 17
Sereikas, Joseph G. 97
707: Year One 19
77 Sunset Strip 18
Shadow of the Thin Man 17
Shark River, FL 29
Sharrar, Paul 138
Sharry *see* Scharry (or Sharry), Sgt. E.
A Shot in the Dark 17
Showboat 70, 79–80
Shreveport, LA 61–62, 102–103, 122, 130–131; *see also* Barksdale Air Force Base
Sibley, Robert Hugh 97

Sicherheitsdienst des Reichsführers-SS (SD) 76
Sierra Nevada 47
SIGMET 25
Silver Dollar 125
Silver Falcon 123
"Simulation and Effects of Severe Turbulence on Jet Airline Pilots" 49
Singapore 72, 106
Singin' in the Rain 17
Sinski, George 11, 14, 15
Sioux Falls Army Air Field, SD 73
Skygods: The Fall of Pan Am 105
Slattery, Ed 160–163
Smigiel, Mrs. Anton 43
Smith, Donald R. 186
Smith, Lynda Gail 97
Smith, Sylvia Lee 97
South America 52, 58, 104, 108, 114
South Farmingdale, NY 65, 96
South Pacific 71
Southwest Airlines 112, 128
Soviet Union *see* Union of Soviet Socialist Republics
Spaulding, Joan Grace 97
Spencer, Kenneth Lee 69–70, 78–81, 97
Splash of Colors: The Self-Destruction of Braniff International 100
Spokane, WA 22, 73
Spring Park, MN 186
squall line(s) 22, 29, 48, 131, 167, 169, 173–175, 177, 183; definition 131
Srodulski, Rose (Mrs. Joseph) 43
SS 76; *see also* Schutzstaffel
SSTs 103, 105–106, 112, 116, 121; *see also* supersonic transports
stabilizer *see* horizontal stabilizer
Stappenbeck, Dr. A.P. 138
Star Trek 18
Starstream 880 *see* Convair 880
Steiner, Roy 20
stick force 87, 93, 95
Stillwater, OK 187
Stinson 100; Detroiter 100
Stuttgart, Germany 74
Sud Aviation 119, 121
Sullivan, Robert 9
supersonic transports (SSTs)

53, 103, 105, 112–113, 116, 121
Switzerland 11

TACAN 24
Tacoma, WA 42, 104
Tamiami Highway 30; *see also* Tamiami Trail or U.S. Highway 41
Tamiami Trail, Old 30
Tampa, FL 104
Taylor, Andy 18
Taylor, Aunt Bee 18
Taylor, Elizabeth 18
Taylor, Robert 79
Tejada, Virginia 186
Tengerstron, E.W. 43
Tengerstron, Mrs. E.W. 43
Texas 44, 59–61, 63–64, 72, 84, 112, 114, 121, 185
The Thin Man Goes Home 17
Thomas, Mary Ann 96
Thorp, Virgil S. 97
351st Bombing Group (Heavy) 73
Thrills, Chills and a Spill 124
thrust reverser assemblies 84
Thunderstorm Project (1949) 172, 174
Time-Life 108
Time magazine 110–111
Tinker, Frank 26
Tinker, Jack 108
Tinker, Jack and Associates 103, 107
Titus, James, 4-C 146
Tobata district, Japan 171
Tokyo, Japan 164, 171
Topeka, KS 138, 146
"Tornado Alley" 2
Toronto, Canada 104
Tracy, Spencer 79
Transcontinental and Western Air 99; *see also* TWA
Trapper John, MD 18
Tripp, Juan T. 105
Truly, Harry Nicholas 97
Tulsa, OK 100–101, 106, 122, 132, 187
turbulence, definition of 25
turbulence penetration speed(s) 38, 62–63, 169, 176–177, 182
Tuscaloosa, AL 72
TWA 99, 123, 169
The Twilight Zone 18
2001: A Space Odyssey 105

Union of Soviet Socialist Republics (USSR) 44, 165

Index

United Airlines 47–49, 51, 53, 55, 63, 91, 99, 100 118, 154, 162; Executive flights 48, 111; flight 227 154; flight 746 47–51, 62, 91; flight 823 162; Mainliners 48
United Kingdom 168
United Nations 70, 77; Commission on the Status of Women 70, 77; Economic and Social Council 77; General Assembly 77
United Press International (UPI) 47, 64, 138, 169; *see also* press pool reports
United States 2, 10, 20, 56, 71, 104, 110–111, 113, 118, 122, 160, 168, 172, 178, 181; armed forces 80; Justice Department 80
United States Air Force 9, 34, 62, 143, 181
United States Army 80; 370th Regimental Combat Team 80
United States Army Air Force 72, 102, 126; Air Defense Command 181
United States Army Air Service 56–57, 73
United States Coast Guard 29, 30, 31, 69
United States Congress 57, 83, 101, 163, 165
U.S. Customs Service 67
U.S. Department of Transportation 117
U.S. Fish and Game Commission 30
U.S. Highway 41 30
United States House of Representatives 165–166
United States National Archives 126
United States National Guard 138, 140, 143–144, 150
U.S. Naval Air Development Center 49
United States Navy 36
U.S. Post Office *see* United States Postal Service
United States Postal Service 56, 57
United States Senate 165
United States War Department 72, 74

U.S. Weather Bureau 22, 25, 36, 41, 45, 66; Aviation *Severe Weather Bulletin 447, Sigmet Bravo 3* 131, 182; Aviation Weather Affairs, 45; *Manual* 25
Universal Aviation Corporation 100
University of Chicago 170, 175
University of Rochester, NY 79; Eastman School of Music 79
University of Wisconsin-Madison 47
University of Wisconsin-Platteville 71
Upper Iowa University 78
US Air 2; flight 1016 2, 178
USA Weekend 12

Valens, Ritchie 10
Van Thullenar, C. F. 34, 35
vertical acceleration 32, 40, 165, 175
Vickers-Armstrong 118
Vickers Vanguard 120–121
Vickers VC10 120–121
Vickers Viscount 118, 120–121, 162
Victoria, TX 186
Vidal, Gore 78
Vietnam 131
Villa Park, IL 97
The Virgin Islands 104
Virginia 90
Volpati, Christian 115
von Goethe, Johann Wolfgang 79
VOR 24
VORTAC 24, 67

Wagon Train 18
Walker, Robert 80
War Department 72, 74
Ward, Charla 159, 186
Washington, D.C. 3, 31, 65, 69, 77, 97, 104, 114, 151, 160–163, 165–166, 186; Dulles International Airport 52, 58, 65, 90, 106; National Airport 161
Waters, Ethel 79
Wauwatosa, WI 97
Wedell-Williams Transport 57
Wells, Dr. H.E. 43
Wells, Mary 108, 110, 112

Welter, Robert D. 186
West Germany *see* Germany, Federal Republic of
Western Air Express 99
Weybridge, UK 120–121
Whipperman, Bernard 188
Whipperman, Mrs. Bernard 188
Whisperliner 123
White House 56–57
Wichita, KS 101
Wichita Falls, TX 100, 122
Wilber, NE 170
Wilber Czech Festival 170
Wilkerson, H.C. 71
Willis, Georgeanna J. 97
Wilmington-Catalina Airlines 99
Wilson, Frank 186
wind shear 2, 66, 174–175
wind-shift line 173–176
wind speed(s) 173–174
Winnipeg, Canada 78
Wisconsin 74
Wisconsin *State Journal* 71, 72
Witness for the Prosecution 17
Witt, Frank 162
Wold Chamberlain Field *see* Minneapolis International Airport
Women's Powder Puff Derby 18
Works Progress Administration 153
World War I 57; Flying Ace 55
World War II 20–21, 58, 70, 75, 77, 82–83, 101, 126, 164, 178
Wright, Donald Keith 186
Wright, Orville 56, 94, 100
Wright engines 53
Wubbold, Joseph 43

Yabakei Gorge 170
yaw damper 55, 129
Yoder, Richard C. 97
Young, Warren R. 45–46
Young, Wilfred (Butch) 144
Younkin, Virginia Lee 24, 42
Yuma, AZ 97
YWCA breakfast 71

Zeng, William B. 64–67, 95–96
Zenkai 170

www.ingramcontent.com/pod-product-compliance
Ingram Content Group UK Ltd.
Pitfield, Milton Keynes, MK11 3LW, UK
UKHW041948140426
5217IPUK00014B/696